The Winters of the World

The Winters of the World

Earth under the Ice Ages

Edited by

BRIAN S. JOHN
MA, D.Phil

DAVID & CHARLES
Newton Abbot London North Pomfret (Vt)

British Library Cataloguing in Publication Data.
The Winters of the World.
1. Glacial epoch
I. John, Brian Stephen
55.17'92 QE697

ISBN 0-7153-7732-9 (hardback)
ISBN 0-7153-7937-2 (paperback)

Set by Northampton Trade Typesetters
and printed in Great Britain
by The Alden Press Ltd
for David & Charles (Publishers) Limited
Brunel House Newton Abbot Devon

Published in the United States of America
by David & Charles Inc
North Pomfret Vermont 05053 USA

Contents

Acknowledgements

We would like to thank the following for kindly permitting the use of copyright material:

Aerofilms: (top) 39; Heather Angel: (top) 111; Hjalmar Barđarson: 187; British Museum: 104; British Petroleum: 229; Chalmers Clapperton: (bottom) 69, 100; E. Derbyshire: 66, 67, 73, (top) 75, (bottom) 76, 81, 82, (top) 83, (top) 85, (top) 88, 90, 92; Department of Energy, Mines and Resources, Canada: 188, 205, 206; R. W. Fairbridge: 135, (bottom) 136, 142, 151; Geodetic Institute, Copenhagen: 10, (bottom) 36, 58, 63, 64, 80, 96, 116, 148, 186, 209, 213; Magnus Johansson: 113; B. S. John: 30, (top) 69, (centre) 75, (top) 76, 86, (bottom) 88, 91, 94, 192; A. Kosiba: (bottom) 75, (bottom) 85; H. A. McClure: 145; J. Ross Mackay: 93; C. Mannerfelt: 89; Mansell Collection: 228; NASA: 115; National Air Photo Library, Canada: (top) 12; David North, 141; Novosti Press Agency: (bottom) 103, 105, 215; Ann Ronan Picture Library: (left) 32; *The Scotsman*, (right) 32; Scott Polar Research Institute: 54; A. J. Smith: 34, (bottom) 165; A. M. Spencer: 122-3; Sven Stridsberg: 126; Studio Jon: (bottom) 87, 112; D. E. Sugden, 68, 181; C. M. W. Swithinbank and the US Dept of Commerce: (bottom) 13, 14; The Tate Gallery, London: 29; S. Thorarinsson: (top) 87, 143; John Topham: 46, (bottom) 111; US Coastguard: (bottom) 12, 53, 97; US Geological Survey: (bottom) 38; US Navy: title page, (top) 36, (top) 51, 98, (bottom) 112,149, 231; Murray Watson/Biofotos: (bottom) 110; Fjellanger Widerøe A/S, Norway: (top) 110; G. M. Young: 117, 127.

We would also like to thank the British Antarctic Survey for the use of photographs taken by BAS personnel.

FACTS

QUATERNARY ICE AGE LASTED ONLY ABOUT
2 MILLION YRS (AFTER B. JOHN)

EARLIER LASTED 50 MILL

CHANGES ENVIRO

EXTIN — DINOS / PLANTS / ADAPT OR VANISH

TIMELINE

Preface

This is a book about the ice ages which have affected planet Earth at intervals throughout the immense span of geological time. At present we are living in an ice age which is commonly referred to as *the* ice age, but it has been under way for so short a time that we cannot be sure about its geological status. So far, the Quaternary ice age has lasted for only two million years or so; yet we know that some of the earlier ice ages lasted for more than 50 million years. The earlier ice ages, like the present one, had profound effects upon the global environment; glaciers covered great parts of the continents, world climates and ocean currents were altered, and plant and animal life was sometimes changed almost beyond recognition. Ice ages have led to large-scale extinctions of plant and animal species; and yet, through the mechanisms of 'glacial stimulus', they have forced innumerable species to adapt and diversify in the face of harsh and rapidly changing climatic conditions. If Man can be said to be a product of the present ice age, many other species must also be the products of earlier ice ages.

The chapters of *The Winters of the World* are organized in a way which should give the reader an impression of what the various ice ages were like. I hope that the book will give an impression of the importance of ice ages in the fashioning of the face of the Earth and in the evolution of our planet's diverse floras and faunas. Chapter 1 sets the scene by demonstrating the global importance of ice and by providing basic geological information which is essential for the understanding of subsequent chapters. Chapter 2 is concerned with the causes of ice ages, and in Chapter 3 Edward Derbyshire examines the effects of large-scale glaciation on the environment. Chapters 4-7 deal with specific ice ages. In Chapter 4 Grant Young examines the very sparse evidence for the ice ages which occurred in Precambrian time. In Chapter 5 Rhodes Fairbridge describes the effects of the Ordovician ice age, paying particular attention to the spectacular discoveries of glacial traces in the Sahara Desert. In Chapter 6 I myself collect together the impressive body of evidence concerning the Permo-Carboniferous ice age, which affected the southern continents in particular.

And in Chapter 7 John Andrews describes the course of the Quaternary —'our'—ice age. This chapter is substantially different from the others; we have so much evidence of Quaternary glaciation and its side-effects that analyses, hypotheses and theories abound in many different fields of Quaternary study. The presentation of some of the modern research findings and speculations may at first sight seem daunting; but we must remember that only by understanding the present ice age can we hope to interpret correctly the events of earlier ones.

Perhaps more to the point, only by discovering the real reasons for the waxing and waning of the Cenozoic icesheets of the middle latitudes will we be able to predict what will happen to world climate over the next few centuries. This is a theme which is taken up again in the final chapter.

The authors who have been invited to write the chapters of this book are all recognized authorities in their own fields.

Edward Derbyshire is Senior Lecturer in Geography at the University of Keele, UK. As a geomorphologist he has made a special study of the impact of glaciation upon the landscape, and he has undertaken field research in many areas including the Arctic and Antarctic.

Grant Young is a geologist who has made a careful study of the evidence of Precambrian glaciation, especially in North America. He is currently on the staff of the Geology Department at the University of Western Ontario in Canada.

Rhodes Fairbridge, Professor of Geology at Columbia University in New York, has long been fascinated by ancient climates. He has written many books and articles on topics related to ice ages, and he has contributed greatly to the continuing debate on theories of climatic change.

John Andrews is currently a professor at the Institute of Arctic and Alpine Research, Boulder, Colorado. He has worked for many years in Arctic Canada on problems related to the growth and decay of the North American icesheet, and he has written stimulating papers on the relations between glaciers and the environment.

Myself? You'll find quite enough about me on the back flap of the jacket of this book.

Inevitably, the chapters of this book reflect the personal interests and opinions of the authors. The discerning reader will find contradictions and discrepancies, but he should bear in mind that the study of ancient ice ages is still in its infancy, and also that, when science is in a debating hall, it is at its best. Many of the theories of modern glaciology and geomorphology have still to be applied to the glaciations of the current ice age, and several years may elapse before they can be applied successfully to the ice ages of the past. I hope that this book will point the way for some routes which might be followed in future research. But, in particular, I hope that the book will bring out the fascination of those times of global catastrophe which we refer to as the 'winters of the world'.

Brian John
January 1979

1

Planet Earth and Its Seasons of Cold

The planet which we inhabit is unusually warm at the moment. We are living in the middle of an ice age, but our global environment is typical of the interglacials which separate the times of extensive ice cover.

Only 18,000 years ago a large part of the northern hemisphere was covered by ice; the great continental icesheets of North America, Scandinavia and western Siberia extended well into the middle latitudes, and there were smaller highland icesheets in tropical latitudes also. The Greenland and Antarctic icesheets, which still exist, were thicker and even more extensive than they are now. Altogether, glacier ice at this time covered about 40 million square kilometres, or almost 30 % of the world's land area.

The effects of the global climatic cooling were not restricted to those areas overridden by glacier ice; the present-day pattern of world climates and vegetation zones was greatly disturbed, and sea temperatures were also lowered drastically. The area of permafrost or permanently frozen ground was expanded until it affected something like 27 million square kilometres (20 % of the world's land area). And at sea floating ice, including pack ice, ice shelves and icebergs, affected approximately 50 % of the world ocean.

Thus, only 18,000 years ago, well over half of the surface of planet Earth was affected by ice of one sort or another, and during spells of exceptionally snowy weather in the northern hemisphere the pervading mantle of whiteness must have covered a surface area much greater still. Visitors to planet Earth at such a time, if they had been familiar with the various forms of water, might well have referred to it as 'Planet Ice'.

The geography of 18,000 years ago was no freak. Ice has affected the surface of our planet on a similar scale on at least four different occasions during the present ice age, and some authorities now believe that there may have been as many as 17 different glaciations during the last 2 million years or so. Some of these glaciations have certainly been more dramatic in their effects than the last one. About 100,000 years ago, for example, the land area affected by glacier ice was almost 45 million square kilometres, and the areas affected by permafrost and floating ice were also correspondingly larger.

The interglacial environment of planet Earth, such as that in which we now find ourselves, is strictly temporary. From the evidence of the earlier glacial and interglacial stages of the present ice age, the periods of cold climate seem to last for about 100,000 years while the periods of warm climate seem to last for less than 20,000 years. The present interglacial has already lasted for more than

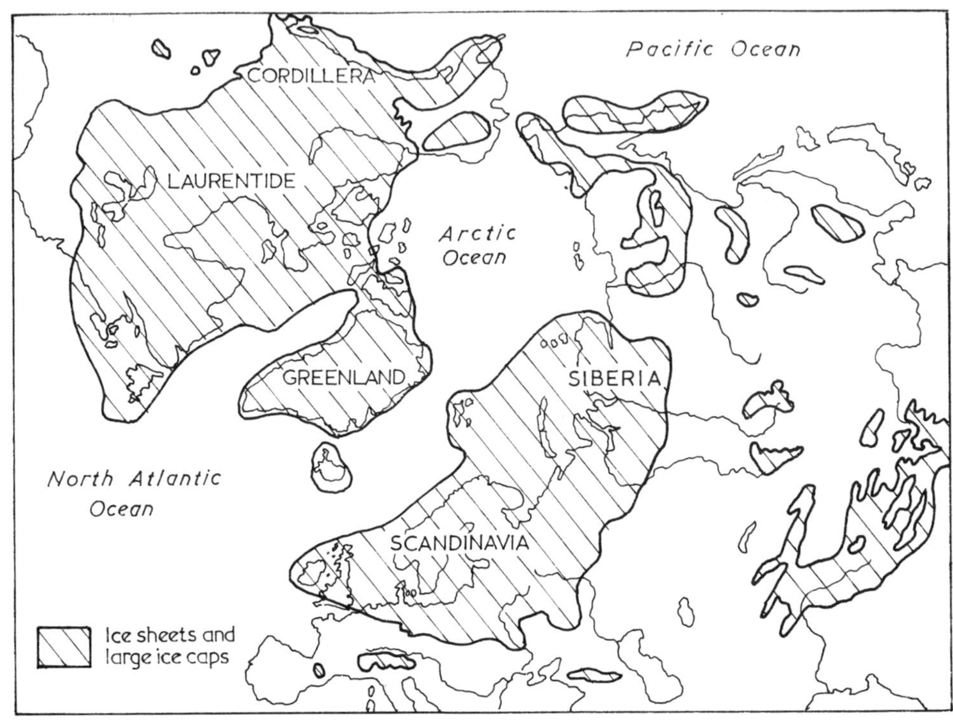

The northern hemisphere icesheets as they appear during a glacial stage. In the southern hemisphere the Antarctic icesheet is normally larger than at present.

The edge of the Greenland icesheet. The area which is buried beneath ice is affected by a number of different processes of glacial erosion and deposition, and beyond the ice edge glacial meltwater is constantly altering the landscape.

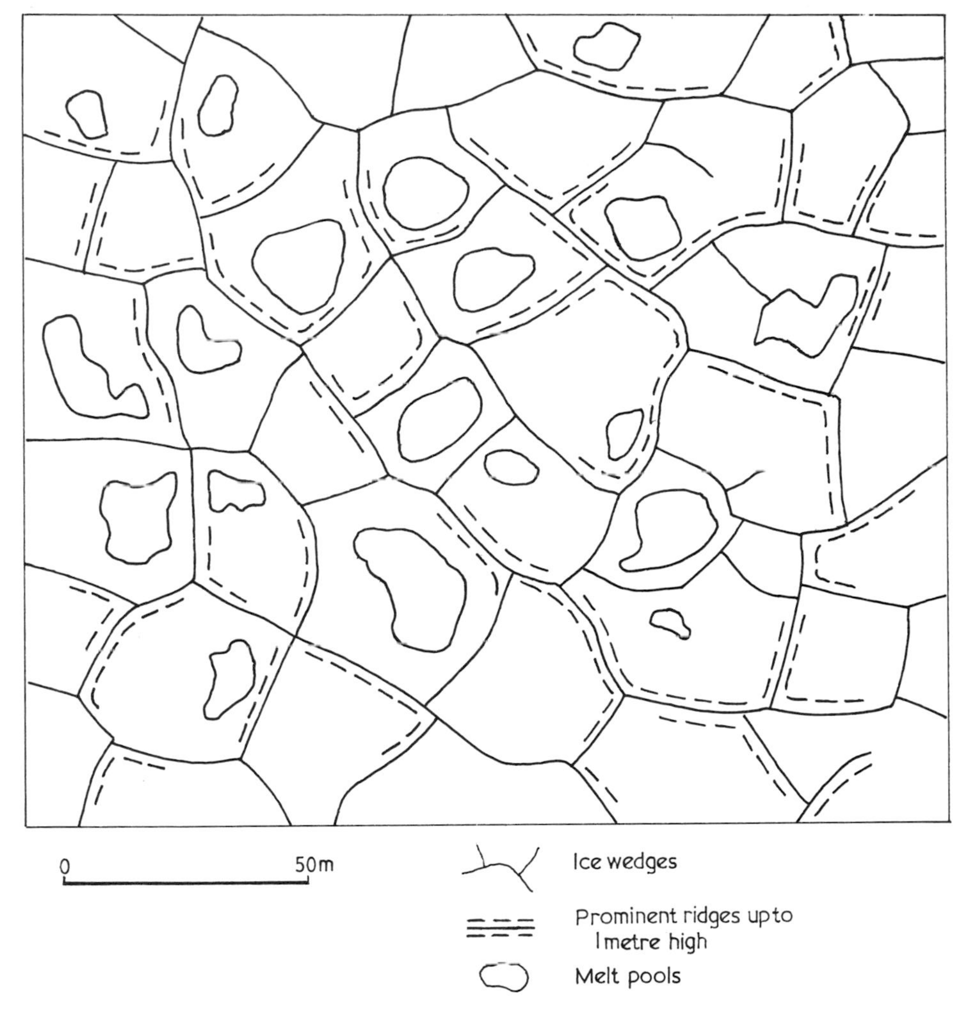

Patterned ground on the Taimyr Peninsula, Siberia. The polygonal markings indicate the presence of permafrost or permanently frozen ground.

half its allotted span, and it is reasonable to assume that from now on the climate will deteriorate gradually until the world's icesheets are once again fully extended.

Even at the present time, close to the warmest part of an interglacial, our planet is affected by ice to an extent which is seldom appreciated. Glaciers cover almost 15 million square kilometres or about 11% of the world's land area. Another 14% of the land area is affected by ice in the ground, and in the northern hemisphere there is a broad belt of country stretching across North America and Eurasia where permafrost is the most important feature of the environment. During the northern-hemisphere winter most of this broad belt, as well as a large part of the middle latitudes, is covered by seasonal snows. At such times more than half of the world's land area is snow-covered.

In addition, large parts of the 'world ocean' are covered with floating ice. The Arctic Ocean is in effect a frozen ocean, and during the winter its surface is covered with about 12 million square kilometres of floating ice. In February

Ice floes in James Bay, Canada. Floating ice of this type must have been widespread in high and middle latitudes during each of the known ice ages.

A small iceberg derived from one of the glaciers of West Greenland and carried southwards towards the coast of Labrador by ocean currents.

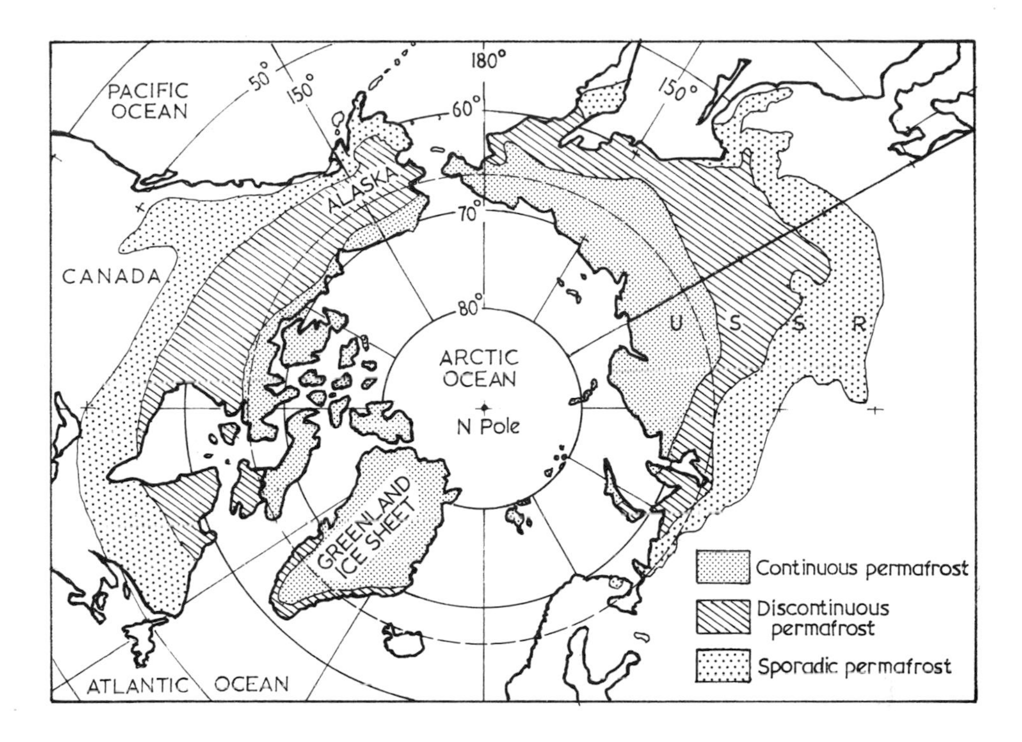

Map of the permafrost belt (area of permanently frozen ground) of the northern hemisphere. Huge areas of Canada and Siberia are affected by permafrost.

Satellite photo of the Arctic regions, in early May 1970, showing the frozen ocean and the main areas affected by snow and ice.

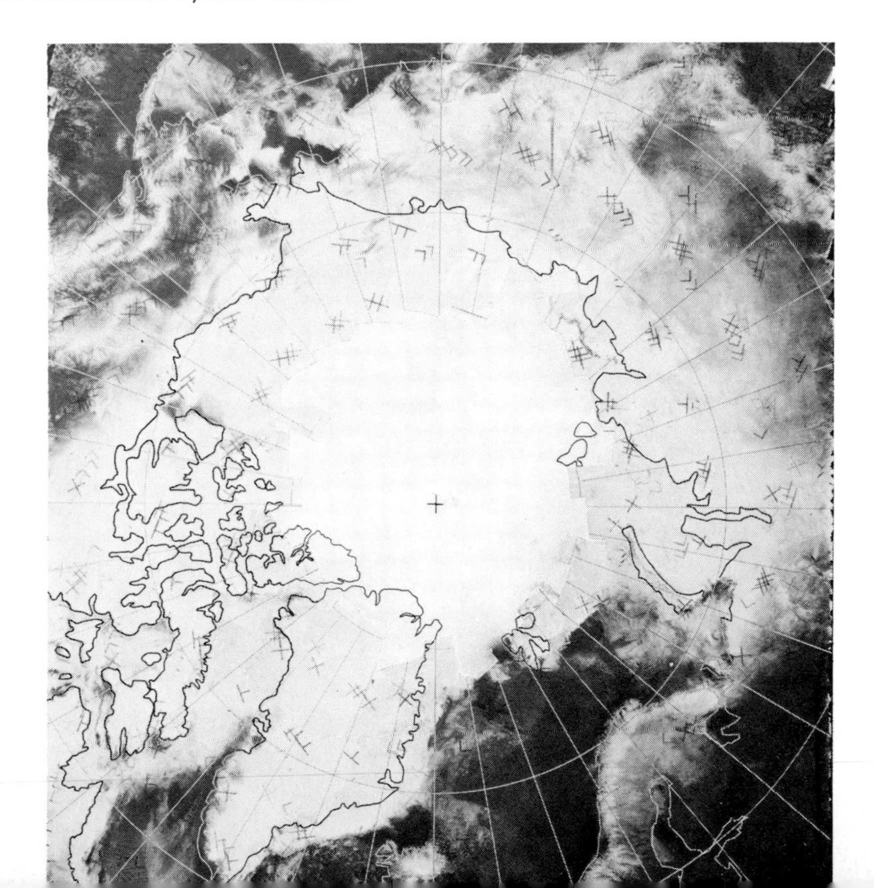

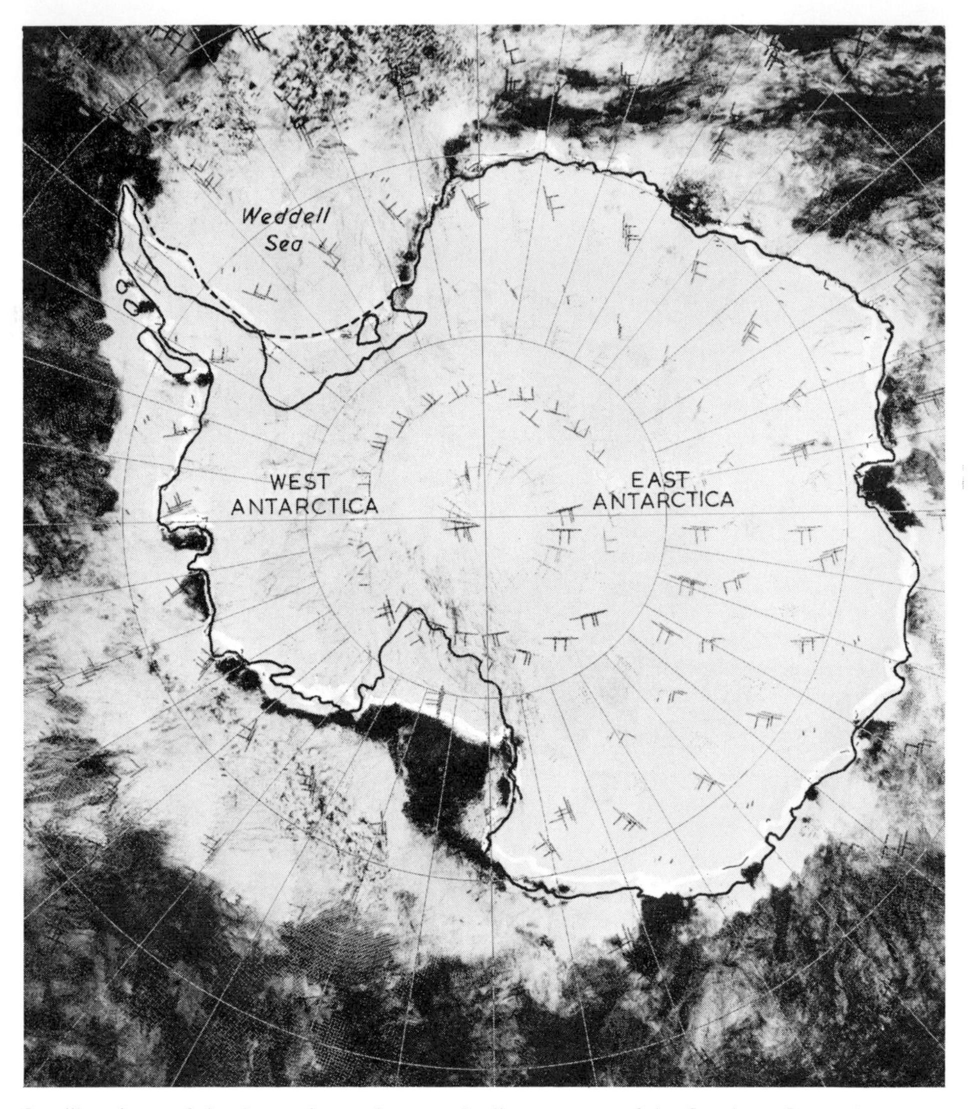

Satellite photo of the Antarctic continent and adjacent parts of the Southern Ocean, in December 1969, showing the Antarctic icesheet and the extent of sea-ice.

and March (the time of the greatest snowcover on the land surface of the northern hemisphere) the sea ice around the coasts of Antarctica is at its least extensive, for this is of course the time of the Antarctic summer. During the Antarctic winter, however, when the Arctic Ocean is enjoying higher temperatures and a seasonal reduction in its cover of sea ice there may be as much as 20 million square kilometres of sea ice in the Southern Ocean around the peripheries of Antarctica. At any one moment about a quarter of the world ocean is affected by floating ice.

The importance of the world's ice cover is gradually being recognized by scientists from many different fields. Something like 75% of all the world's fresh water is stored in glacier ice, and most of this is locked up in the Antarctic icesheet. The world's great ice masses in the polar regions have a profound effect upon the workings of the global 'climate machine', and they also control the oscillations of sea-level. Glaciers and ice-covered areas of the oceans are being studied more and more intensively, since they contain many clues concerning both the processes involved in environmental change and the spacing of climatic changes, which occur on a number of different scales. Only a decade ago it was customary to refer to the global environment as made up of four main 'spheres': the atmosphere, or envelope of air; the lithosphere, or layer of rock and sediment on the Earth's surface; the hydrosphere, or water layer; and the biosphere, or sphere of living things. Nowadays it is becoming customary for the writers of textbooks in environmental sciences to refer to a fifth sphere—the cryosphere, or sphere of frozen water. Not before time, the vast importance of ice on planet Earth is becoming recognized.

The above paragraphs refer to the global importance of ice during the current ice age. But they could equally well refer to the glacial and interglacial conditions which prevailed during the Permo-Carboniferous ice age, or the late Ordovician ice age, or the Varangian ice age, all of which are dealt with in later chapters of this book.

The winters of the world have occurred intermittently throughout geological time; it is the purpose of this book to summarize what we know about them and to analyze their importance in an evolving world inhabited by an ever more diverse succession of lifeforms.

It is often said that Man is a product of the present ice age. Could it also be the case that earlier ice ages have stimulated plant and animal evolution in a similar way? This is but one of the questions which this book attempts to answer.

A Sense of Time

Before we can obtain a realistic impression of the spacing of ice ages and of their importance for the evolution of living things, it is necessary to gain some understanding of geological time. During the eighteenth century it was still commonly believed that the total span of geological time was of the order of 6,000 years. Later, a number of scientists estimated that it would have taken at least 20 or 30 million years for the known thicknesses of sedimentary rocks around the world to have accumulated, and even the estimates based upon other criteria seldom placed the age of the Earth at more than 40 million years. Only with the widespread use of radiometric dating during the middle part of the present century has a reliable 'absolute' timescale for our planet been established.

It is now widely accepted that the age of the Earth is about 4,600 million years. Although there are no known rocks of this age on Earth, meteorites which are assumed to have formed at about the same time have been dated as 4,600 million years old, and so have the oldest rocks from the Moon. The oldest known rocks on Earth, collected in Greenland, are about 3,800 million years old. These are from the earlier part of the Precambrian 'eon', the first of the major divisions of geological time.

The geological timescale is still difficult for most people to comprehend. A small child waiting for his dinner thinks of 30 minutes as 'a long time', and even adults who have been taught to think of time in a reasonably realistic way find it difficult to imagine the conditions which prevailed on the surface of the Earth more than a century ago. In spite of his accumulated wisdom, Man is still incredibly naïve in his appreciation of time. Yet, in order to understand the evolution of our home planet, we have to try to understand the span of geological eons, eras, periods and epochs.

The Precambrian eon accounts for about 85% of all the time which has elapsed since the formation of the Earth. We still know remarkably little about the 4,000 million years or so of Precambrian time, but gradually geologists are beginning to break this immensely long eon down into smaller time units. About 600 million years ago the first organisms apart from single-celled ones began to appear, and during the Palaeozoic era there was a gradual evolution of lifeforms. About 440 million years ago the first land plants appeared, and animal life took to the land about 45 million years later.

During the Mesozoic era the continental landmasses began to move towards their present positions on the surface of the globe, and there were rapid changes of climate. Dinosaurs appeared, only to become extinct at the end of the Cretaceous period.

During the following Cenozoic era, occupying the last 65 million years or so, the climate has gradually deteriorated, culminating in the Quaternary ice age. The Quaternary period has also seen the emergence of *Homo sapiens*, modern Man. It is a salutory experience to realize that Man, who frequently thinks of himself as *the* child of Mother Earth, has been present for only a miniscule part of geological time. If we think of geological time as spanning some 46 'years', Man is approximately four *hours* old.

The Spacing of Ice Ages

Ice ages have occurred through geological time at fairly regular intervals. Although there are differences in the estimated durations of the various ice ages, the later ones seem to recur at intervals of approximately 150 million years. The reasons for this spacing are obscure, but the problem of causes is discussed in detail in Chapter 2.

Table 1 The Spacing of Ice Ages Through Geological Time

Name of Ice Age	*Approx. Age*	*Approx. Duration*	*Geological Period*
Cenozoic	1 million years	10 million years (so far)	Quaternary and Tertiary
Mesozoic?	150 million years?	Unknown	Jurassic?
Permo-Carboniferous	300 million years	50 million years?	Permian and Carboniferous
Late Ordovician	450 million years	25 million years?	Silurian and Ordovician
Varangian or Eocambrian	600 million years	20 million years?	Upper Proterozoic
Sturtian or Infracambrian I	750 million years	50 million years?	Upper Proterozoic
Gnejsö or Infracambrian II	900 million years	50 million years?	Upper and Middle Proterozoic
Huronian (probably two or three ice ages?)	2,300 million years	200 million years?	Lower Proterozoic

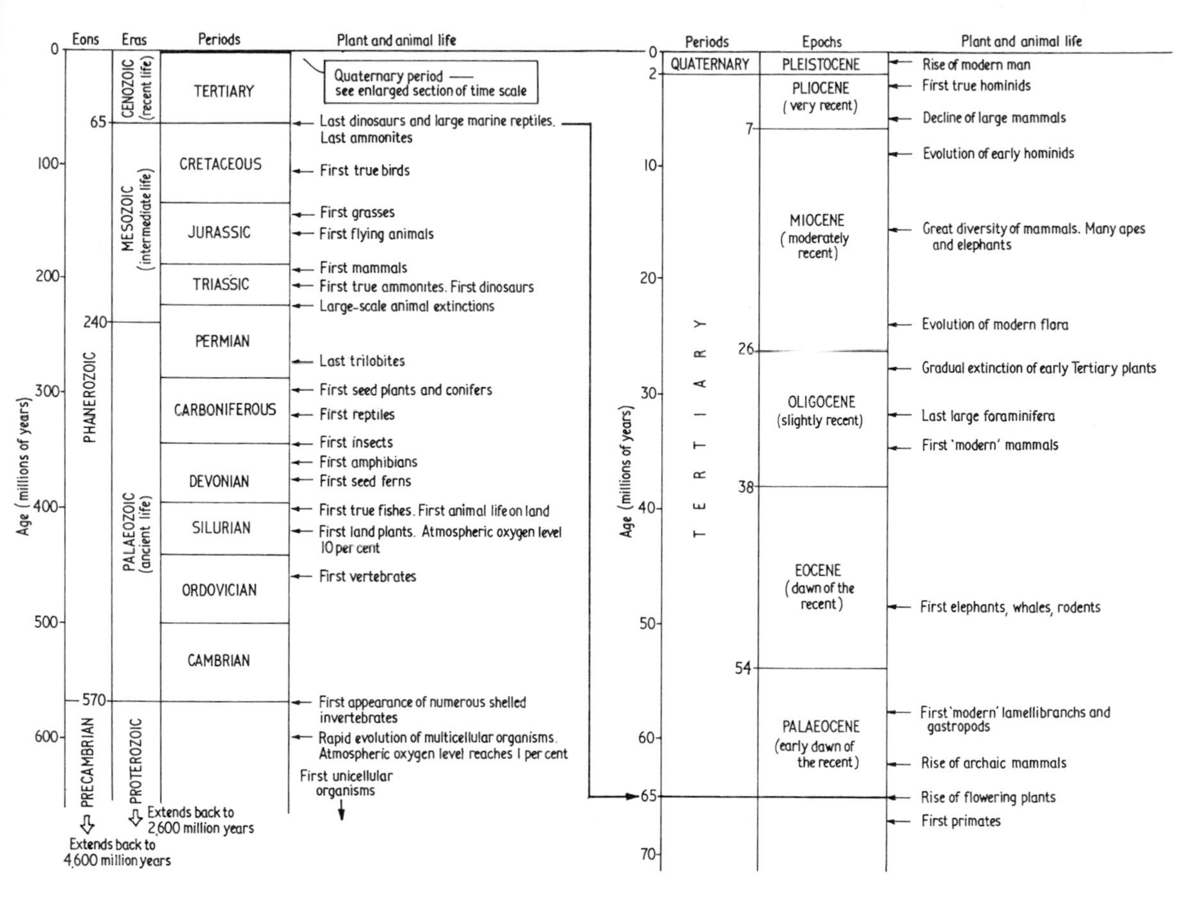

The geological timescale, based for the most part upon radiometric age determinations. The right-hand column shows some of the main 'events' in the evolution of plants and animals.

There is one notable anomaly in Table 1; where one might expect widespread evidence of an ice age about 150 million years ago, during the later part of the Jurassic period, there are only a few small traces of glacial action in the geological record. Perhaps this means that the apparently regular spacing of the other ice ages is the result of chance? On the other hand it may simply mean that, although some environmental conditions were favourable for the onset of a Jurassic ice age, the circumstances were not quite right.

While there may have been a slight cooling of global climate at this time, there was no large landmass in the vicinity of either the south pole or the north pole. In addition, large parts of the continents were submerged beneath shallow seas, and this would have inhibited the growth of icesheets. A final important point concerns the lack of a world-wide orogeny (period of mountain-building) during Jurassic time. Each of the other ice ages recorded in the upper part of the geological column is related to a great period of mountain-building and, although it is difficult to separate cause and effect, there must be some link between widespread land uplift and widespread glaciation. This whole question is examined in greater detail in the next chapter.

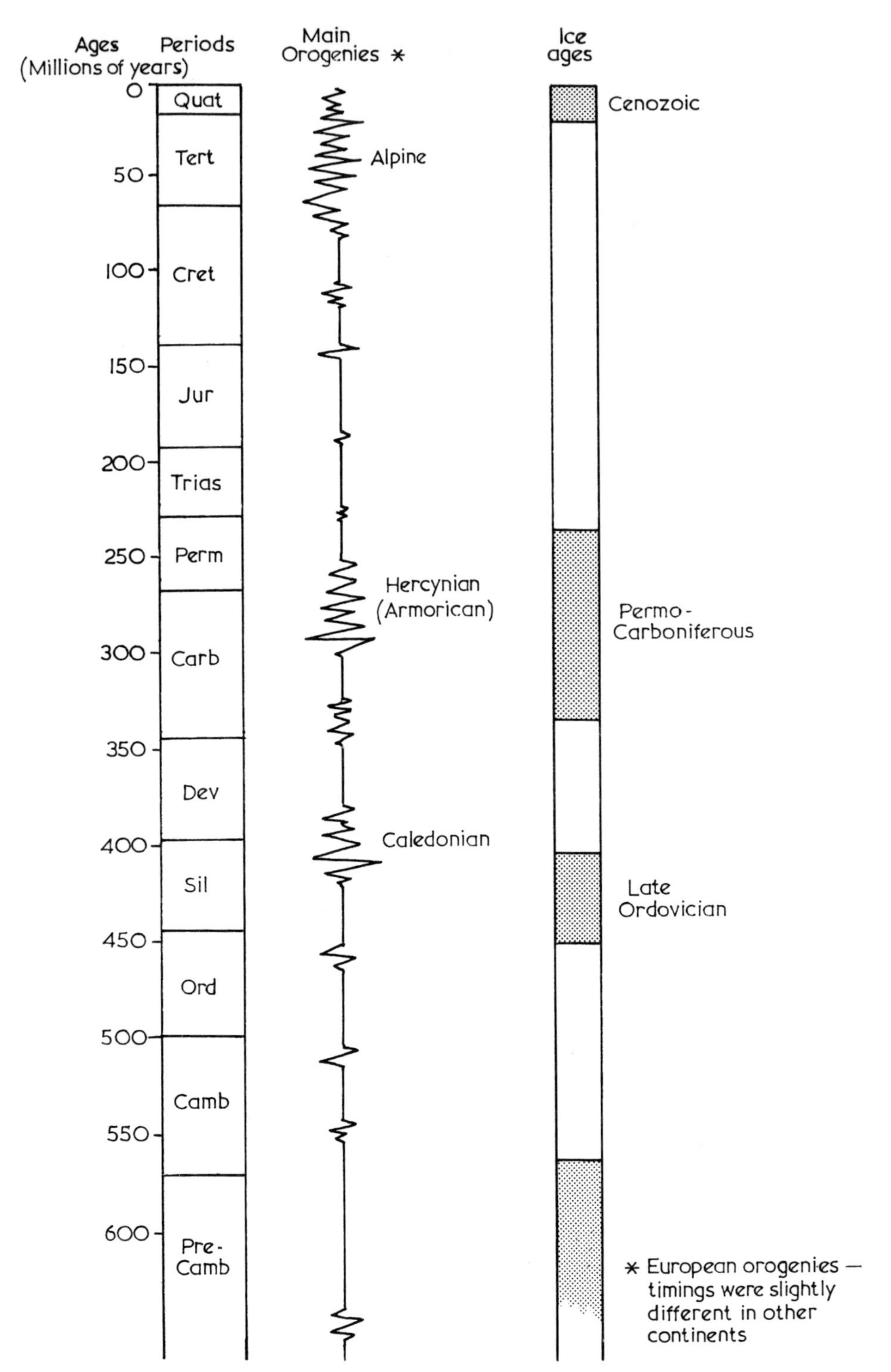

The spacing of periods of mountain-building and ice ages during the last 600 million years. The main orogenies occurred at different times on the various landmasses, although a number of particularly pronounced and widespread mountain-building episodes can be discerned on all continents.

The Geological Basis

From the foregoing, it will be apparent that an understanding of ancient ice ages depends to a large extent upon an understanding of some of the ideas of modern geology. The following brief outline may therefore be found useful by those who are unfamiliar with the developments which have transformed the Earth sciences during the last 20 years or so.

The surface of the Earth is made up of solid rock materials which are referred to as the crust. Beneath the oceans the crust is normally about 6km thick, whereas it is usually 30-40km thick beneath the continents. These two types of crust, referred to as 'oceanic' and 'continental', are different in composition and density, but silicate minerals rich in alumina are plentiful in both.

The next layer beneath the crust is the mantle, thought to be about 2,900km thick and composed very largely of solid (but nevertheless mobile) material with silicate minerals rich in iron and magnesium. The upper part of the mantle, together with the crust, forms a rigid layer called the lithosphere.

The central layer or core of the Earth is very dense and largely liquid, consisting of a mixture of nickel and iron.

Although the crust and mantle are generally referred to as solid, they are not perfectly rigid. Under certain conditions they deform rather like toffee, pitch or ice: that is, they flow very slowly when force is exerted gradually, but shatter when force is exerted rapidly. These two properties are commonly seen in rocks exposed at the ground surface which may be either folded or faulted. The folding or bending of rock strata (layers) is usually associated with gradual pressure exerted from beneath; if the pressure builds up to the point where the rock cannot deform any further it has to crack or split, and rock layers are forced apart along the resulting fault plane. We can see small-scale evidence of these processes every time we look at a sea cliff made up of deformed rock layers, but the buckling and splitting of rock also occurs on a vast scale. Both the crust and the mantle are moving all the time very slowly under the action of the various forces which operate inside the Earth, and this *mobility* is of fundamental importance for any understanding of the distribution and appearance of continents and oceans, mountain-chains and plateaux, basins and coastal lowlands.

The flow of the mantle causes huge pieces of the lithosphere to slide horizontally over the Earth's surface, thus in time bringing areas of unrelated rocks into contact with one another. These 'rafts' of material are now called plates, and the structural processes associated with their movements are referred to as plate tectonics. There are six major plates and six or seven minor ones, and together these plates cover the whole of the Earth's surface. They are composed of both crustal and mantle material welded together.

Where adjacent plates are in contact with one another different types of boundaries have been recognized. Where two plates are moving apart, as in the centre of the Atlantic Ocean, basaltic material from the mantle wells up to the surface to create new oceanic crust and lithosphere. As this material is added to the divergent plates, they are dragged apart by convection currents in the mantle which act like giant conveyor belts. The process is called sea-floor spreading, and it leads to the creation of the pronounced mid-ocean ridges which have now been mapped in all of the deep ocean basins. Sea-floor spreading

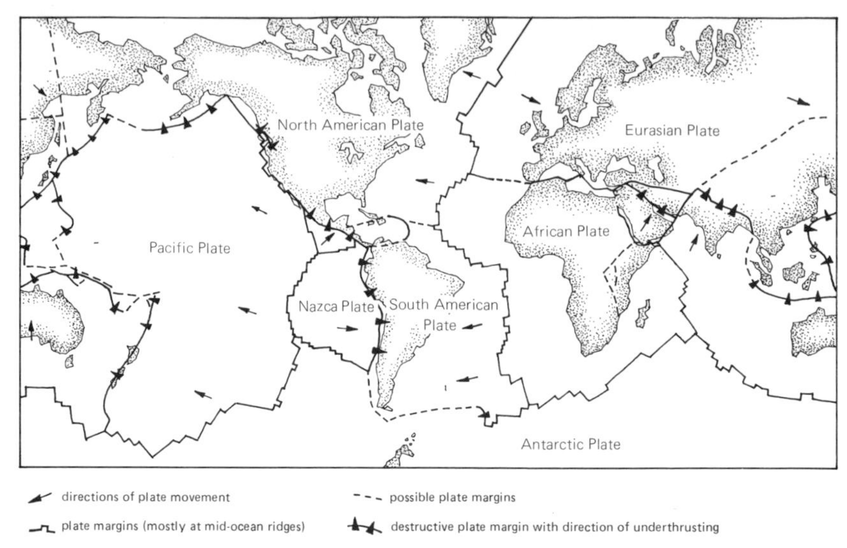

Map of the main plates of the world today, showing also the positions of mid-ocean ridges and mountain ranges affected by plate collisions.

also leads to the widening of oceans and the separation of the continents.

Where two plates are in collision there is always a destructive margin, marked by the subduction of the material of one plate, which is carried down into the mantle whence it came millions of years earlier, beneath the other. The subducted material is composed usually of dense oceanic crust, whereas the lighter and more buoyant continental crust is deformed but not consumed. 'Active' margins of this type are usually associated with large-scale disturbances of the geological environment, such as the creation of deep-sea trenches, the intense folding of 'new' sediments, the uplift of mountains, volcanic activity and earthquakes. On boundaries where two plates are sliding past one another there may be faulting and earthquakes, but no dramatic creation of mountains or ocean trenches.

According to the modern theories of sea-floor spreading and plate tectonics, there is no difficulty in explaining the distribution of active volcanoes and earthquakes on our planet; both occur for the most part on or adjacent to the boundaries between plates, either where new crust is being created or where old crust is being consumed by subduction. The maps of current volcanic activity and recent earthquakes 'match up' to an extraordinary degree with the maps of plate margins.

Nor is there any difficulty in explaining the distribution of the extreme low and high points on planet Earth. Deep-sea trenches are formed on active continental margins where one plate margin is being carried down beneath another. The 'downward drag' of the crust in the subduction zone is largely responsible for the deepening of the trench, but there is also a sideways squeezing effect which causes great deformation of the sediments which accumulate in the trench. Deep depressions full of sediments are normally called geosynclines; geologists now recognize several different types, but in essence all of them form at, or very close to, colliding plate margins. Mountain chains are also created for the most part along these collision zones as a result of the compression of continental plate margins. The forces involved in plate collisions are quite adequate

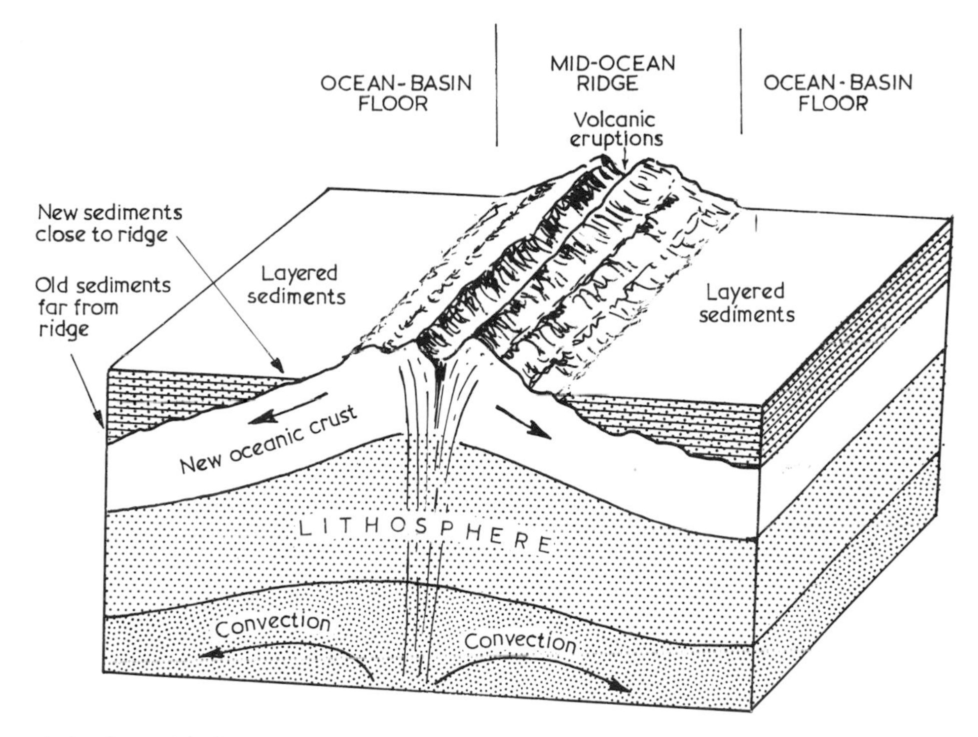

A simple model of a mid-oceanic ridge, showing how new crust is created. New sediments, as they accumulate on the ocean floor in the vicinity of the ridge, are transported away as if on a conveyor belt.

A model of the zone of contact between two colliding plates. Oceanic crust is being subducted beneath continental crust, leading to the deepening and compression of a deep-sea trench and the building of new mountains on the continental margin.

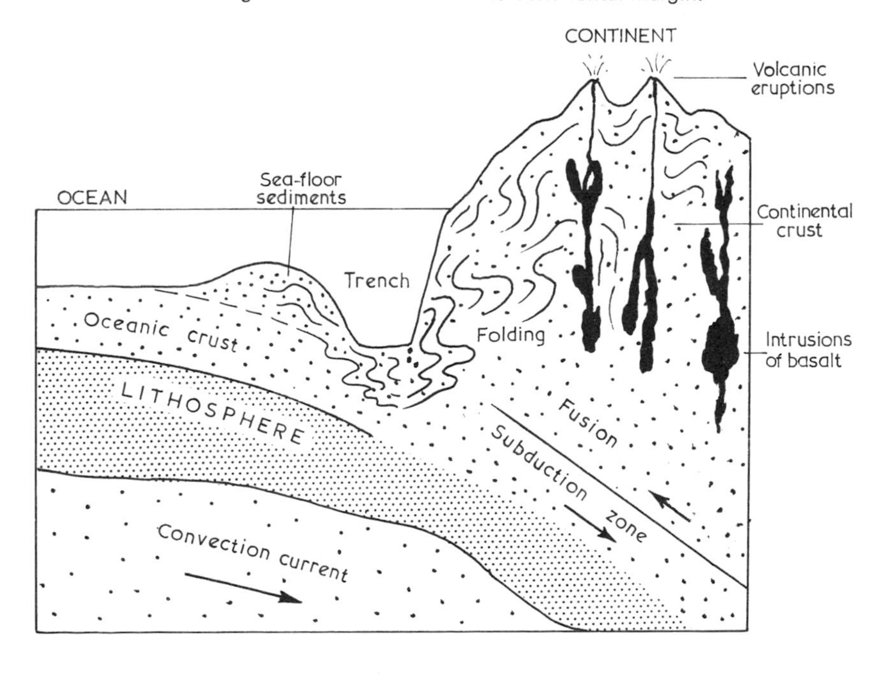

to compress and shatter great thicknesses of old continental rocks and to uplift mountain ranges such as the Rockies, Andes, Himalayas and Alps to great altitudes. The distortion of old rocks is usually accompanied by intrusive and extrusive volcanic activity and by widespread faulting accompanied by earthquakes. Some mountain areas are created from the compressed sediments which fill ocean trenches, and each phase of mountain-building, or orogenesis, leads to a complicated 'mixing' of old and new continental rocks. On rare occasions, blocks of oceanic crust are also incorporated into mountain chains as they are built. These mountain chains may be largely above or largely below sea-level.

Away from the plate margins, there is much greater stability of the crust, and both in the oceans and on land there are ancient plains, plateaux, denuded mountains and infilled basins which are no longer greatly affected by tectonic forces. In the continental interiors there are immensely old cratons; for the most part these are composed of Precambrian crystalline rocks (in the so-called shield areas) but there may also be platforms of old sedimentary and metamorphic rocks with traces of domes, arches and basins where there were once geosynclines. Many of the rocks in the continental interiors are more than 600 million years old.

On the ocean floors, the huge smooth areas away from the mid-ocean ridges are called abyssal plains. These are the areas where deep-sea deposits are gradually accumulating. Eventually these deposits will be transported by the 'conveyor belt' mechanism of sea-floor spreading towards the plate edges, where they will be brought into contact with the sedimentary rocks formed as a result of the slow but steady denudation of the continents. While the oceanic crust is 're-cycled' by subduction, the rock debris carried seawards by the erosion of the continents by water, wind and ice is returned to the continental landmasses by the process of mountain-building.

The Mobile Continents Through Time

It follows from the above summary that the arrangement of continents and oceans is anything but fixed. The present arrangement of landmasses has evolved only during the past 100 million years or so, since the Cretaceous period. Before the Cretaceous, and indeed throughout geological time as a whole, continents have been created, split apart and reassembled on many occasions. Sometimes there have been vast 'supercontinents' stretching almost from pole to pole and spanning many degrees of longitude; on other occasions the landmasses have been more fragmented, with a scatter of continents, oceans and seas widely dispersed across the face of the Earth. Where continents have been split apart, seas and oceans have evolved to occupy the rifts in the crust; where continents have come into contact, either through head-on collision or through sliding past one another, mountain ranges have been created. In some areas, segments of the sea floor have been buried and consumed by the mobile crust; in other areas, the sea-bed sediments have been compressed and buckled and 'used' for the creation of mountains. Some new mountains remain largely beneath the surface of the sea, with their tips projecting as jagged islands. Some mountains, such as the Himalayas and Rockies, are uplifted thousands of metres above sea-level.

Twenty-five years ago these ideas would have been largely dismissed as 'catastrophist', even although a brilliant visionary called Alfred Wegener had assembled a convincing body of evidence to support them as early as 1912. For half a century Wegener's ideas were heavily criticized, often with considerable venom, or just ignored. Then, in the mid-1960s, new and exciting discoveries in the field of geophysics caused Wegener's much-criticized and sadly neglected ideas to be reassessed. Suddenly the theory of continental drift became acceptable, and it formed the basis for the theories of plate tectonics and sea-floor spreading which have revolutionized geology and geophysics. At last there was a comprehensive theory which allowed the scattered observations of more than a century from many branches of geology to be set into a consistent and intelligible framework.

It is not the purpose of this chapter to examine the ideas of continental drift and plate tectonics in detail. However, if we wish to understand the ice ages which have punctuated geological time, we must familiarize ourselves with at least some of the changes which have occurred in world geography over the past 2,000 million years or so. For example, we cannot understand the Permo-Carboniferous world and its ice age by referring to a map showing the modern arrangement of continents and oceans. Maps are now available which show the 'palaeogeography' of Permo-Carboniferous times, and of the other periods of Earth history in which we are interested. These maps have been constructed on the basis of evidence from many fields, such as stratigraphy, palaeontology, tectonics, palaeoclimatology, vulcanology, geomagnetic studies and geomorphology. In many respects they are inaccurate, but they are the documents on which all future geology textbooks will be based. They are also basic for the understanding of the chapters which make up this book.

The five maps we show here depict the changing positions of the continents as far as they can be interpreted from the evidence currently available. These maps can be referred to during the reading of Chapters 4-7, but it will also help if the following brief points are kept in mind:

Precambrian: 600 million years ago. The map shows that some of the continents may have been recognizable in their outlines even 600 million years ago. The widely dispersed pattern of continents shown is that which followed the

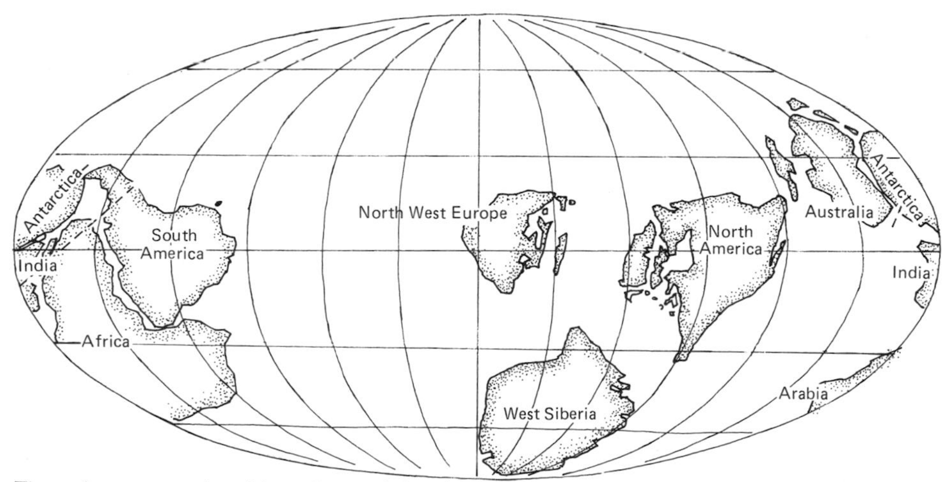

The palaeogeography of Late Precambrian time.

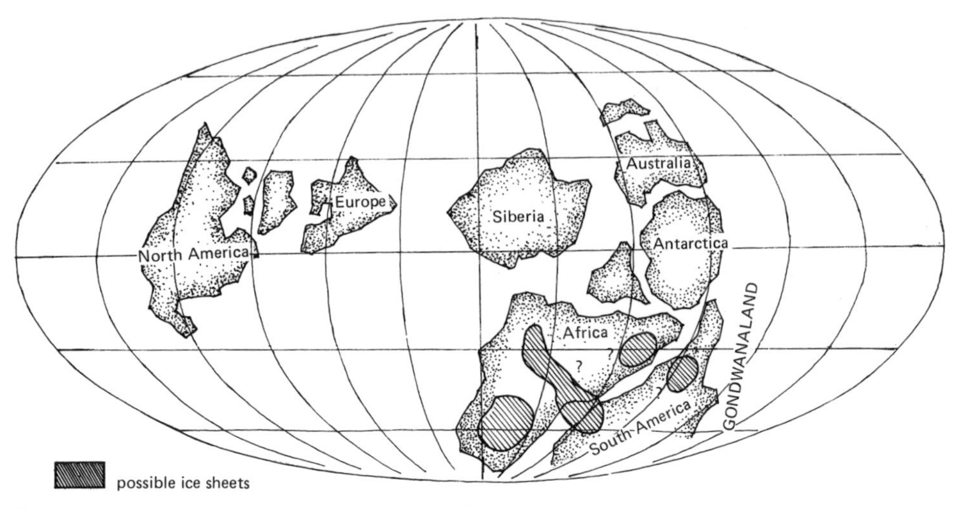

The palaeogeography of Late Ordovician time.

break-up of the 'Proterozoic Supercontinent' which existed from about 2,200 million years ago to about 1,200 million years ago. This supercontinent was interesting in that Africa and Arabia were at that time located between South America and North America. While the supercontinent was in existence there occurred at least two prolonged ice ages, usually referred to together as the Lower Proterozoic or Huronian Glaciation. After the break-up of the supercontinent there was another ice age or series of ice ages which affected every continent—except, ironically it may seem, Antarctica. This time of cold has been referred to as 'the most extensive period of glaciation in Earth history', for it spanned a period of almost 600 million years. It is worth noting that this is the equivalent of the whole of Phanerozoic time from the beginning of the Cambrian period to the present day. Bearing in mind the points made on page 16 concerning the periodicity of ice ages, one may doubt whether any *single* ice age could have lasted for 600 million years. A *series* of ice ages is perhaps more likely, and indeed some authorities have recognized in the field evidence and in the radiometric dates a grouping of *three* ice ages referred to as the Genjsö, Sturtian and Varangian ice ages. The matter of Precambrian ice ages is discussed more fully by Grant Young in Chapter 4, but at this stage it is worth noting that the 'recurrence interval' of about 150 million years between one ice age and the next may be just as relevant for the Precambrian as for the later part of geological time.

Late Ordovician: 450 million years ago. By late Ordovician times there had been considerable movement of the continental plates across the surface of the globe. In the west the ancient equivalents of North America and Eurasia were isolated in a vast ocean and separated from one another by an early 'Atlantic Sea'. In the east there was a grouping of continents into a supercontinent which is generally referred to as Gondwanaland. This included the old stable 'blocks' of Africa, South America, India, Antarctica and Australia. To our eyes Africa and South America appear to be 'upside down', and the late Ordovician south pole was located in what is now the Sahara. Australia was located in the middle latitudes of the northern hemisphere, and the Siberian landmass was isolated

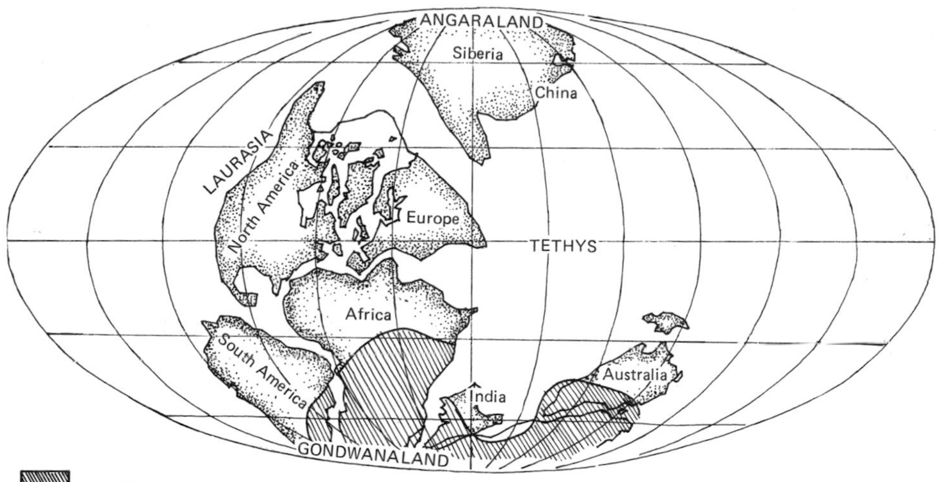

The palaeogeography of Late Carboniferous time.

from all the others by a deep marine gulf. With this arrangement of continents in mind, we should not be surprised at the idea of an Ordovician glaciation of the Sahara Desert, discussed by Rhodes Fairbridge in Chapter 5.

Late Carboniferous: 300 million years ago. By the end of the Carboniferous period there had been further great changes in world palaeogeography. There were now three large continents: Laurasia, made up of today's North America, Greenland and Europe west of the Urals; Angaraland, made up of China and Siberia; and Gondwanaland, extending across South America, Africa, Indai, Antarctica and Australia. Laurasia occupied the equatorial latitudes, while Angaraland had its centre at about 60°N. Gondwanaland at this time had the greater part of its land area in the high latitudes of the southern hemisphere, and this explains the presence of a number of different ice centres in late Carboniferous time. As explained by myself in Chapter 6, the continental icesheet of Gondwanaland probably grew from the coalescence of many smaller icesheets, and large ice masses also persisted during the early part of the succeeding Permian period.

Jurassic: 150 million years ago. During Jurassic times the continents began to take up more familiar positions. During the earlier Triassic period the main landmasses had all been joined together in one huge supercontinent, usually referred to as Pangaea. The two segments of this supercontinent, Laurasia (in the north) and Gondwanaland (in the south), had been ranged around three sides of a great ocean which we refer to as the Tethys. Now, during the Jurassic, Pangaea began to split apart. A wide gulf was opened up between the northern supercontinent of Laurasia and the southern supercontinent of Gondwanaland. Large parts of Laurasia were inundated by a rise in sea-level, and Angaraland was separated from North America. There was also a split in Gondwanaland as South America and Africa were separated from India, Antarctica and Australia. The South Atlantic and North Atlantic oceans had still not been formed. No evidence of widespread glaciation has been found for Jurassic time. Both the north pole and the south pole were located in oceanic areas. The land area with the highest latitude at this time was east Antarctica, and if evidence of

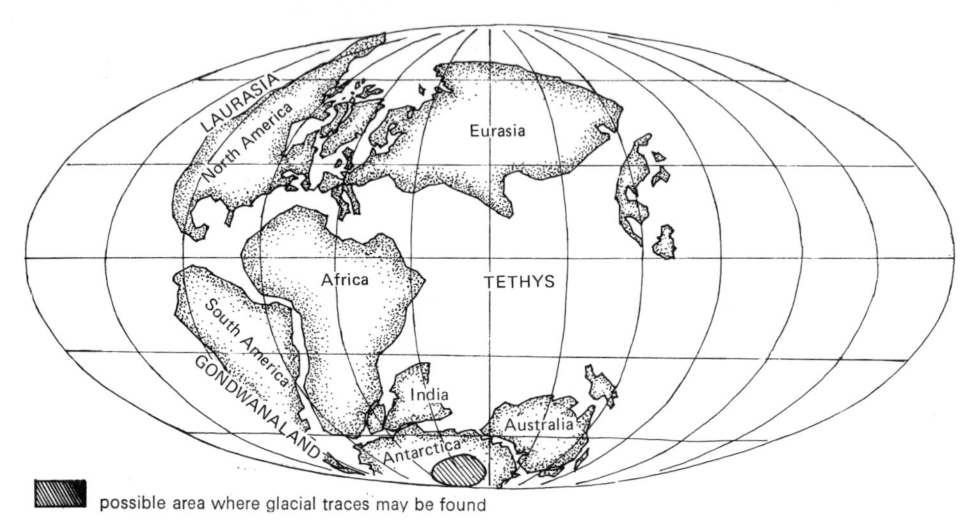

The palaeogeography of Jurassic time. There are no signs of extensive glaciation in the known geological record.

Jurassic glaciation is to be found it will probably be contained in the Jurassic rock sequence of the area which is at present submerged beneath the Antarctic icesheet. One Jurassic tillite has already been discovered in Antarctica, but its extent is unknown.

Quaternary: one million years ago. By the beginning of the Quarternary period the continents had taken up approximately their present positions on the surface of the globe. Within the last 100 million years or so North America and South America have drifted away from Eurasia and Africa, and they have also been linked together during the latter part of this time by the Isthmus of Panama. Antarctica has remained in its polar position while Africa, India and Australia have migrated northwards by varying amounts. Whereas the Pacific and Indian Oceans are of great antiquity, it is only in the last 100 million years or so that the Earth has seen the creation of the North and South Atlantic, the Arctic and the Southern Oceans. The supercontinent of Afro-Asia is also a 'new' geological feature. The climatic cooling of the Cenozoic era has culminated in the Quaternary ice age, with periodic expansions of great icesheets over North America, Scandinavia and Western Siberia, parts of the Arctic Ocean, eastern Siberia and Central Asia. Two 'persistent' icesheets have remained in position even during the interglacial stages of the ice age. Even a brief look at the world map of the present day reveals an ideal situation for an ice age; there are large continental areas in both hemispheres which are close enough to the poles to permit icesheets to grow and survive.

The Evolution of Life

While it would be foolish to pretend that all of the main evolutionary 'events' on planet Earth have been connected with ice ages, it can be argued that the winters of the world have had a considerable impact upon both plant and animal life. For example, during the later part of Precambrian time there was apparently a rapid evolution of multicellular organisms, with the appearance of mineralized skeletons which were capable of being preserved as fossils. Within a few million years, brachiopods, sponges, gastropods and trilobites evolved and colonized

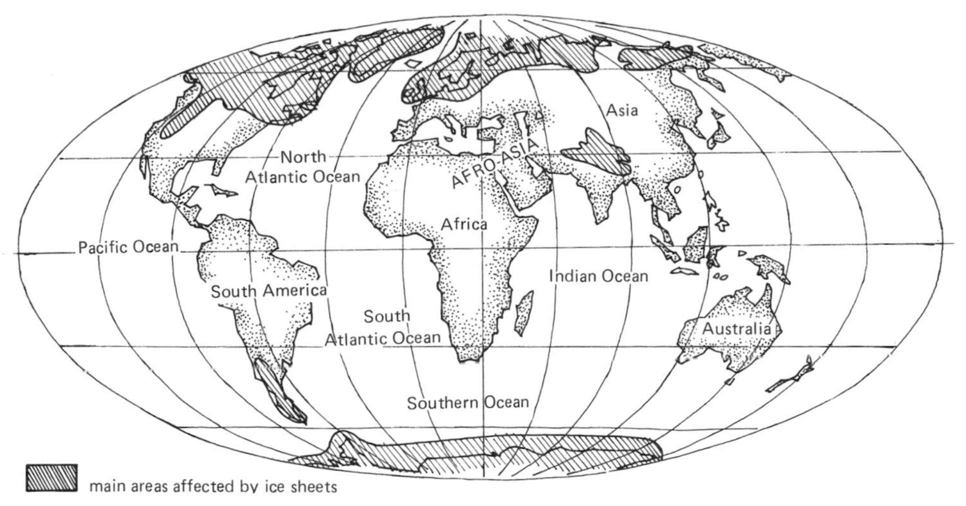

The geography of the Quaternary world, showing the extent of the icesheets during a typical glacial stage.

large parts of the world ocean. This evolutionary 'explosion' may have been linked with the build-up of free oxygen in the atmosphere; but could it also have been linked with the stimulus of rapid environmental changes during one or more ice ages? During the Ordovician period, the time of the next ice age, the cooling of world climate led to the evolution of climatic zones more or less as we know them today. These zones were more highly compressed than they were before, since an ice age is always associated with a sharp increase in the temperature difference between polar and equatorial regions. Marine animals were forced to specialize, and in the cold waters of the polar seas new species evolved. As the numbers of trilobites declined, so many new genera of brachiopods appeared to occupy new environmental 'niches'. During the Silurian period, following the end of the ice age, there was an 'explosion of life', with many classes of both plants and animals moving ashore to colonize the continents. About 300 million years ago, during the Permo-Carboniferous ice age, there appeared the *Glossopteris* flora of Gondwanaland. This flora seems to have evolved rapidly in response to the climatic cooling of the ice age, and it dominated the vegetation of the southern continents for over 100 million years. During the Jurassic period there was another great episode of floral and faunal diversification, followed during the present ice age by the spectacular diversification and evolution of mammals. Within the space of the last 10 million years or so the gradual expansion of the Cenozoic ice cover and the periodic compression of climatic belts have been accompanied by the extinction and redistribution of flora and fauna on a vast scale. Finally, under the stimulus of rapid climatic change during the Quaternary period in particular, Man has evolved to dominate all of the other plant and animal species of the planet.

The foregoing paragraph is composed of but a few disconnected threads from the complicated weave of evolution. Very many other key events in the history of evolution have been identified from the warmer seasons of planet Earth, and these cannot of course be explained by reference to the stimulus of ice ages. And yet it remains true that most of the textbook discussions of evolution devote little or no space to the factor of 'glacial stimulus'. Most of the evidence for the evolution of plant and animal species through geological time comes from locations which were once in tropical latitudes, where climatic changes are 'damped down' because of the relative stability of the equatorial climate even during ice ages. Hence most of the explanations of evolution are concerned not with climate but with biological factors, competition for food supply, cosmic radiation, the composition of the atmosphere, and so forth. Essentially these explanations have little to do with either palaeogeography or palaeoclimatology, and indeed it is often assumed that the global climate, once established, has remained rather stable for the past 600 million years or so.

The winters of the world show us that the climate has been anything but stable through geological time. This book demonstrates that the planet's intermittent ice ages have been linked with spectacular climatic and geographic changes, leading on at least seven different occasions to drastic alterations in the character of the global environments available for plants and animals. Each ice age must be accompanied by the following:

1. Widespread glaciation by continental icesheets in favourable locations in the high and middle latitudes.
2. Widespread 'alpine glaciation' in the high mountain areas of all latitudes.
3. The build-up of ice shelves in favourable environments in polar latitudes.
4. A great extension of floating ice in high latitudes, with mobile floes of sea ice and icebergs affecting many millions of square kilometres of the world ocean.
5. The development of widespread permafrost in many high-latitude and mid-latitude areas not affected by icesheets or smaller glaciers.
6. Substantial changes in atmospheric circulation, with increased temperature gradients in the middle latitudes, increased storminess over the oceans and desiccation in the continental interiors of the tropics.
7. Alterations to the pattern of oceanic circulation, with ocean currents blocked and diverted as a result of the growth of icesheets.
8. Rapid changes of sea-level of at least 250m as expanding and contracting icesheets periodically extract and return water which normally resides in the oceans.
9. Violent changes in the positioning and extent of vegetation belts, with consequent great changes in the distribution of animals.

These direct and indirect effects of glaciation, seen in the context of geological time, are dramatic if not catastrophic. They deserve to be considered seriously by anyone who wishes to understand the history of planet Earth and the evolution of its lifeforms.

BRIAN JOHN

2
Ice Ages: A Search for Reasons

An imaginative reconstruction of Noah's Flood by Francis Danby (1793-1861). During the seventeenth and eighteenth centuries the Flood was widely believed to have been responsible for the fashioning of the landscape.

The idea that glaciers were once far more extensive than at present dates only from the middle of the last century. Before 1840 the conventional view of the surface of the Earth was much coloured by the Biblical creation myth and by the idea that Noah's Flood was responsible for the majority of landforms. In 1654 Dr John Lightfoot, the learned vice-chancellor of Cambridge University, stated that 'Heaven and Earth . . . and clouds full of water and Man were created by the Trinity on 26th October 4004 BC at nine o'clock in the morning'. As late as 1750 it was still heresy in the Christian world to claim that the Earth was more than some 6,000 years old, and it was still widely believed that the land surface bore abundant evidence of the forty days and forty nights of Noah's Flood. For example, deep valleys and lake basins in the mountains could be explained as the products of erosion on a gigantic scale by a deluge of water, and many of the surface deposits of northern Europe could be interpreted as flood deposits. Various sediments such as boulder-clay and bedded sands and gravels were referred to as 'drift', and this term has remained in use in Western

Europe until the present day. Erratic boulders scattered about the landscape, often in improbable positions on hilltops and perched on cliff edges, could not be explained by any easily observable process, and so they too were seen as evidence of the power of flood water during the deluge.

These catastrophist ideas caused considerable unease among certain scientists. About 1500 Leonardo da Vinci realized that not all landforms could be explained in biblical terms, and during the eighteenth century James Hutton developed and expounded what came to be known as the 'Law of Uniformitarianism'. In essence, this law states both that the past is the key to the present and that the present is the key to the past. Hutton argued that the surface of the Earth is undergoing constant change, and that all changes in detail can be explained in terms of natural, observable and commonplace causes. His ideas were in direct conflict with the popular catastrophist notions of the time, and inevitably there was great opposition to them. However, after the publication of his influential *Theory of the Earth* in 1795, and after considerable support in the writings of John Playfair and Sir Charles Lyell, opposition was crumbling rapidly by the mid-1830s. Now it was widely accepted that running water was responsible for the formation of river valleys, and that rivers organized themselves in a manner which allowed the continuous operation of the processes of erosion, transport and deposition.

The time was ripe for a realization that glacier ice was of great importance in the shaping of the land surface. In 1795 Hutton expressed the opinion that the Alps had formerly been covered by a great mass of ice, with granite blocks and other erratic boulders carried for considerable distances to the Jura mountains and elsewhere by immense glaciers which had since disappeared. Nobody listened, in spite of the fact that the peasants of the Alpine valleys accepted all this as self-evident. In 1821 a Swiss civil engineer named Ignaz Venetz-Sitten repeated Hutton's ideas, and in 1824 Jens Esmark proposed, independently, that the glaciers of Norway had once been much more extensive. In 1829

A typical erratic boulder transported by glacier ice and left on a bare ice-scratched rock surface. Before the advent of the glacial theory, such boulders were thought to have been carried by the flood waters of the Deluge.

Venetz stated his belief that the plains north of the Alps, and indeed the whole of Northern Europe, had been affected by ice, and in 1834 Jean de Charpentier read a paper strongly supporting Venetz and providing convincing proof from field observations.

After this the Glacial Theory attracted an increasing number of proponents. Among the most influential was Louis Agassiz, who talked of the work of ice in a famous address to the Helvetic Society in 1837. He was also the first person convincingly to demonstrate the former existence of glaciers over wide areas of the UK and North America.

William Buckland, professor of geology at Oxford and previously one of the greatest proponents of the submergence concept, became convinced of the value of the Glacial Theory in the interpretation of British scenery, and it was he who invited Agassiz to visit Britain in 1840. On October of that year *The Scotsman* newspaper shocked the non-scientific world by reporting that Scotland had once been covered by glacier ice. In British scientific circles the last effective resistance to the ideas of Agassiz was destroyed by two classic papers, one published in 1862 by T. F. Jamieson and the other in 1863 by Archibald Geikie. In the USA the arrival of Agassiz in 1846 did much to speed up the wide acceptance of the Glacial Theory, and J. D. Dana's influential *Manual of Geology* (1863) strongly supported many of the ideas of the glacialists. At last Earth science was relieved of the burden of Noah's Flood.

The concept of ice ages is of course related to the Glacial Theory. Here the pioneer was a forestry professor called Bernhardi, who wrote in the Heidelberg *Jahrbuch für Mineralogie* for 1832 of a time when glacier ice had extended all the way from the north polar regions to cover parts of Germany. In 1837 Karl Schimper built upon this imaginative idea and referred to a glacial period or 'Eiszeit' in Switzerland, and in the same year Agassiz, in his famous address, talked of a 'great ice period' caused by climatic changes and marked by a vast sheet of ice extending from the North Pole to the Alps and even into central Asia. Originally he believed that this 'ice period' had occurred *before* the uplift of the Alps; it was not until 1847 that he realized that the Alps were present before the onset of glaciation and that the former glaciers of northern Europe were quite separate from the Alpine glaciers.

Gradually the idea of an ice age gained acceptance. During the 1850s both A. C. Ramsay in the UK and A. Marlot in Switzerland recognized that there had been distinct cold and warm phases during the Quaternary period, and this idea was supported over and over again in the decades which followed. As the science of geology developed, stratigraphic studies began to reveal that the warmings and coolings of climate had been matched by periodic advances and retreats of the glaciers. The later part of the nineteenth century saw a continuing debate on both sides of the Atlantic on the definition of the Quaternary period and the Pleistocene epoch, the dating of the boundary between the Pliocene and the Pleistocene, and the number of glacial and interglacial stages which could be recognized in the stratigraphy and fossil content of sediments both within and beyond the margins of the glaciated areas. During the 1890s the foundations of the glacial-stratigraphic column of central North America were laid by James Geikie, T. C. Chamberlain and T. Leverett; and the year 1909 saw the publication of one of the most influential Earth science texts of all time. This was *Die*

DISCOVERY OF THE FORMER EXISTENCE OF GLACIERS IN SCOTLAND, ESPECIALLY IN THE HIGHLANDS, BY PROFESSOR AGASSIZ.

Professor Agassiz of Neufchatel has recently been studying the glaciers of the Alps with great care, and has been led by his investigations to certain bold, novel, and highly interesting conclusions respecting the part which these singular bodies have acted in the physical history of the globe. We were present when he explained his views in the Geological Section of the British Association at Glasgow, and subjoin a short outline of them, drawn up, we believe, under his own eye. To this we are enabled to add a most interesting communication, addressed by him to Professor Jameson, bringing out the important fact, that *he has discovered distinct indications of the ancient existence of glaciers at Ben Nevis, and elsewhere in the Highlands.* There are other geological phenomena which lead us to conclude that the climate of this country at a former, but, geologically speaking, a recent epoch, was much colder than it now is; and that such a state of things is consistent with the course of nature, is shown by the fact that Mr Darwin found glaciers reaching down to the level of the sea on the west coast of Chili, in latitude 46 degrees, that is, 11 degrees nearer the equator than Ben Nevis. The evidence on which M. Agassiz's experience in the Alps has taught him to rely, consists, first, of the striated and smoothed surfaces of the rocks, the direction

Louis Agassiz, one of the most influential of the 'glacialists', who were responsible for the gradual acceptance of the 'glacial theory' during the mid-nineteenth century.

Part of the article in *The Scotsman* newspaper for 7 October 1840, which announced that Scotland had been affected by glacier ice.

Alpen im Eiszeitalter ('The Alps during Glacial Periods') by Albrecht Penck and Edward Brückner. Using the evidence of soil profiles, river-terrace studies and data on rates of weathering, they proposed that the northern Alpine region had been affected by four main glacial stages and three intervening interglacials. The glacials were named, from oldest to youngest, Günz, Mindel, Riss and Würm. This scheme remains fundamental to our understanding of the Quarternary period even today, in spite of the fact that it is a gross simplification.

*Table 2 The Classical Glacial Sequences of Central North America and the Northern Parts of the European Alps**

North America	*European Alps*	*Climate*
Wisconsin	Würm	Glacial
Sangamon	Riss-Würm	Interglacial
Illinoian	Riss	Glacial
Yarmouth	Mindel/Riss	Interglacial
Kansan	Mindel	Glacial
Aftonian	Günz-Mindel	Interglacial
Nebraskan	Günz	Glacial
	Donau-Günz[2]	Interglacial
	Donau[1]	Glacial

* It should no longer be assumed that the North American and Alpine sequences are exactly time-equivalent. Some modern correlations 'match up' the glacials and interglacials in quite a different way.

[1, 2] Stages not originally recognized by Penck and Brückner.

The idea of pre-Quarternary ice ages is almost as old as the Glacial Theory itself. In the 1850s W. T. Blandford discovered ancient glacial deposits in the Permo-Carboniferous rocks of tropical central India, and in the following decades traces of the work of ice were discovered in the rocks of South Australia, Natal in South Africa, and southern Brazil.

By the turn of the century there were so many reports of such occurrences from rocks of all ages and in all climatic zones that Earth scientists were beginning to accept the hypothesis of intermittent global cooling throughout geological time. The parallel hypothesis of intermittent glaciation in appropriate areas has become more and more widely accepted, and the 'winters of the world' must now be acknowledged as periods of profound importance for our understanding of the global environment.

The Causes of Ice Ages

A worldwide ice age cannot commence until two basic requirements have been satisfied. In order for large ice masses to grow and survive in previously unglaciated areas there must be a global lowering of temperatures so that snowfall becomes one of the main forms of precipitation and so that winters snowfall are not melted away each summer. Also, there must be an adequate moisture supply over a long period to provide the raw material out of which glacier ice is made.

Both of these requirements appear quite simple, but when they are examined in detail they raise innumerable questions. The most basic of these questions concern the mechanisms of global cooling and prolonged periods of snowfall. *How* are temperatures lowered over a large part of the globe? Must the basic cause be external, related to astronomical or atmospheric factors? Or has planet Earth got some sort of internal instability which allows global temperatures to drop as part of a chain reaction or feedback mechanism? Could global cooling follow some apparently quite innocuous small-scale event in the atmosphere? Or does the global climate machine require some geological or atmospheric catastrophe to set the cooling cycle in motion? The rest of this chapter is concerned with some of the causes and 'trigger mechanisms' which have been proposed by Earth scientists, climatologists and even astronomers as responsible for the ice ages which have intermittently affected our planet. Some of the theories mentioned below are discussed again in Chapters 4-7, where they are relevant for the understanding of particular ice ages.

Internally Generated Causes

1. *Crustal wandering*. The term 'crustal wandering' was coined by Professor Frederick Shotton to describe the changes which have occurred through geological time in the arrangement and latitude of the various landmasses on the face of the Earth. Since the adoption of Alfred Wegener's ideas on continental drift, it has become widely accepted that both the supercontinents and various detached landmasses have been moving about over the Earth's surface, transported like rafts afloat on the fluid mantle, ever since the early part of the Precambrian eon. As shown in the previous chapter, the present arrangement of continental plates is strictly temporary; it is quite unlike that of 150 million years ago, and quite unlike that predicted for 150 million years' time.

When large landmasses are located in middle- or high-latitude areas where the

A rock pavement affected by an ancient glacier at Irai, Central India. This type of evidence convinced some nineteenth-century geologists of the reality of ancient ice ages.

climate is cold enough for large-scale glaciation, icesheets can build up. The extent of the glaciation may then be increased by climatic feedback mechanisms, with the area of cold climate (and hence glacial build-up) affecting mid-latitude areas also. Icesheet glaciation will continue so long as a continental landmass is in a 'convenient' location for low temperatures and adequate moisture supply; when it drifts away into equatorial latitudes (or, in some instances, into a polar location where there is not enough snowfall to maintain an icesheet), glaciation comes to an end.

It is worth mentioning at this point that the most convenient location for the build-up of an icesheet is not necessarily *polar*. Polar regions are certainly cold, but they also have a tendency to be arid. As we see from the unglaciated areas of Alaska, North Greenland and Siberia today, cold dry regions close to the poles are quite unsuitable for the build-up of large glaciers. On the other hand, the mountain territory further south, warmer and wetter, is much better suited to glacier growth.

Another aspect of crustal wandering is the tendency of the poles themselves to shift over geological time, usually as a result of temporary or longer-term 'wobbles' in the direction of the axis of the Earth. It is known that the poles have changed their positions during earlier geological epochs, and one theory postulates that the present ice age commenced soon after the Earth's poles took up their present position at the beginning of the Pleistocene. Naturally a shift in the position of the poles would alter the latitudes of landmasses independently of the operation of plate tectonics, bringing some areas into convenient locations for icesheets to develop.

2. *Mountain-building*. When a land surface is elevated the effect is to increase the area above the snow-line. There may also be a disproportionately large effect upon regional climate, with reduced air temperatures and increased precipitation helping to extend the snow cover beyond the mountains themselves. Under such circumstances, glaciers will almost certainly develop.

It is now widely accepted that the present ice age commenced as a result of a

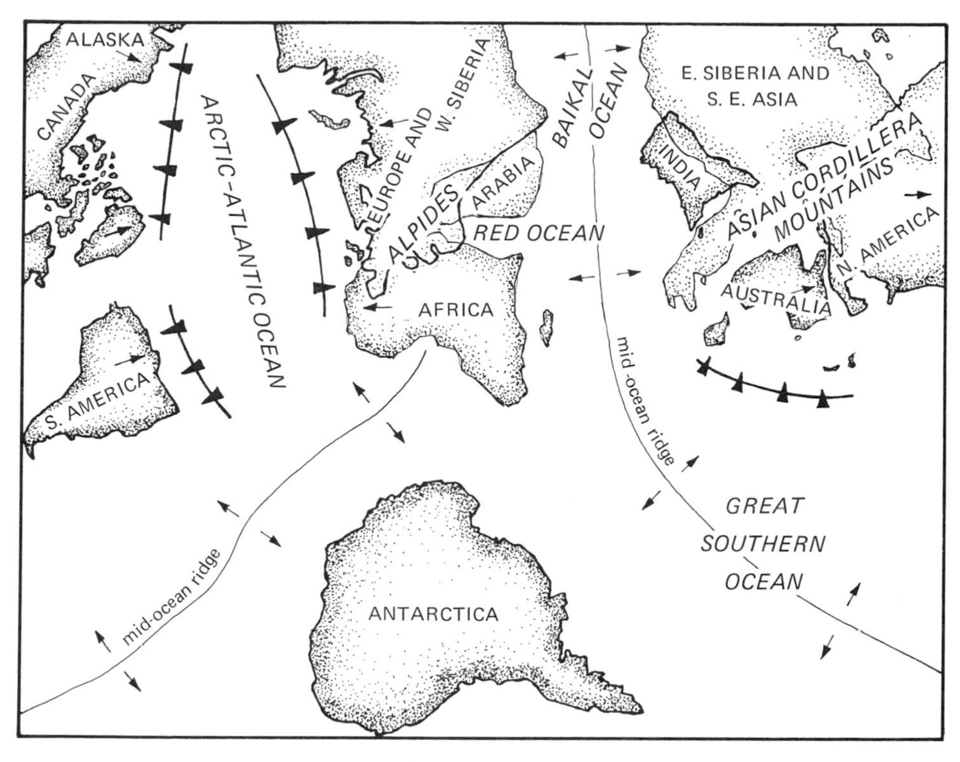

The predicted arrangement of continental plates in 150 million years' time.

global cooling of climate during the Tertiary, which was in turn affected by a phase of widespread continental uplift. In particular, many new mountain areas were created by tectonic forces as the continental landmasses took up approximately their present positions on the Earth's surface. The Alps were uplifted by some 2,000m from the Pliocene to the mid-Pleistocene; over the same period the Himalayas were raised 3,000m; and during the Pleistocene alone the Sierra Nevada range of the USA appears to have experienced uplift of more than 2,000m. In Chapter 1 it was suggested that the onset of earlier ice ages may also be connected with periods of mountain-building; for example, the Permo-Carboniferous ice age succeeded a period of large-scale uplift and orogenesis in many parts of the world.

During a period of mountain-building the main centres of glaciation are initially the mountains themselves, with widespread highland ice caps, valley glaciers and snowfields. The glaciers then extend into the lowlands, where piedmont glaciers, ice caps and eventually icesheets develop. So long as the mountain range remains at high altitude, and so long as it remains at the same latitude, this type of glaciation is perpetuated by abundant snowfall in the uplands and on lowland glacier surfaces. However, if the mountains are eventually worn down beneath the level of the regional snow-line, or if the snowfall is reduced (maybe by the creation of new land where previously there was ocean, causing a reduction in the moisture supply carried by onshore winds), then glaciation comes to an end.

Rugged scenery near the coast of Alaska. Much of this area has never been greatly affected by glacier ice, since it is too arid for large glaciers to develop.

Mountains affected by high snowfields and glaciers in East Greenland. Abundant snowfall ensures the growth and survival of glaciers.

3. *The expanding Earth.* Some authorities believe that the gradual fall of sea-level, which is well documented for the Tertiary period, is at least partly the result of a slow growth in the circumference of the Earth. According to this theory, as the Earth's surface area increases, the water in the oceanic reservoir has to spread itself more thinly, leading to an overall fall of sea-level and an increase in the land area. Coastlines retreat seawards, perhaps by many hundreds of kilometres where there are wide continental shelves, increasing the climatic 'continentality' of many landmasses and cooling down those regions no longer affected by oceanic influences. Also, with larger land areas the brightness (or albedo) of the Earth's surface is increased, leading to a loss of heat from incoming solar radiation which would otherwise be absorbed by the oceans. World climate might be cooled sufficiently for the initiation of an ice age such as that in which we now find ourselves.

It is doubtful whether the expanding Earth theory is relevant for the earlier part of geological time, for it does not explain the many marine *transgressions* which have punctuated the geological record.

4. *Supercontinents and sea-level.* Many geologists and palaeontologists now believe that the creation and splitting of supercontinents are of great importance for sea-level rises and falls of the order of 500m or more. A current theory states that, when ocean-floor spreading is going on (i.e., when supercontinents are being fragmented or split apart), the emergent and volcanically active ridges have such a large volume that they form shallow seas in the centres of the ocean basins. The result is that oceanic waters are displaced, causing marine transgressions around the edges of the continental landmasses.

Conversely, when a supercontinent is 'assembled' there are few mid-oceanic ridges being created. The old ones subside, causing an increase in the volume of the ocean basins with a consequent marine regression or sea-level fall. The latter condition might lead to the onset of an ice age, for the cooling of an enlarged supercontinent, particularly if it is located in the middle and high latitudes, could be enough to ensure the build-up of ice caps and icesheets.

The Permo-Carboniferous ice age coincided with the 'assembly' of the supercontinent of Pangaea. On the other hand, the late Precambrian ice ages coincided with the fragmentation of an early supercontinent, and the Ordovician and Cenozoic ice ages can also be interpreted as occurring at times of wide continental dispersal. The 'supercontinent theory' is clearly not adequate on its own to account for the growth of an extensive ice cover.

5. *Variations in the Earth's core.* A number of authors have suggested that surface changes on the Earth may be related to internal changes—for example, the output of heat from radioactive materials inside the Earth or alterations in the direction and strength of the convection currents which may exist within the mantle. These changes could affect the geothermal heat flux (the rate at which heat is transmitted through the Earth's crust to the surface), the strength or frequency of volcanic eruptions and earthquakes, and also the Earth's magnetic field.

The magnetic field is now widely accepted as having some relationship with global climate, and it has been suggested that some magnetic 'reversals'

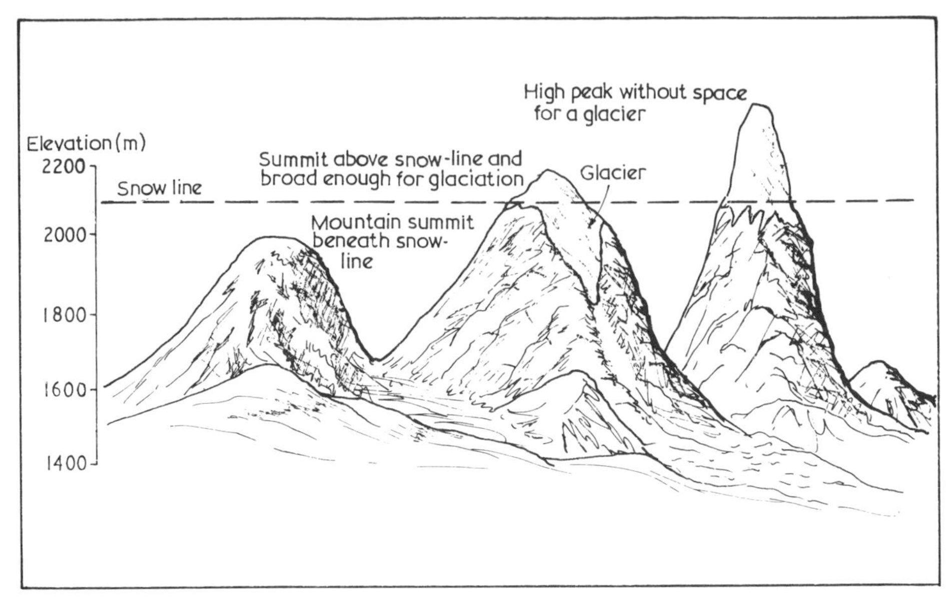

The snowline in a mountain area is the line above which snowbanks may persist throughout the year. If a large part of a mountain lies above the snowline, glaciers can develop.

Part of the Rocky Mountain chain in Alaska. These are 'youthful' mountains elevated above the snowline in recent geological times and still affected by earth-movements. Part of the mountain peak in the centre of the photograph fell away during the 1964 Alaskan earthquake.

Signs of a falling Cenozoic sea-level. Ramsey Island, off the coast of Wales, is typical of many areas on the oceanic peripheries in having a sea-floor 'platform' well above present sea-level. The rocky hills must have been small islands when the platform was being eroded by the sea.

Coincidence of global ice ages and periods of marine regression. It is difficult to be certain that there is a cause/effect relationship, since there are many other regressions recorded in the geological column which did not coincide with ice ages.

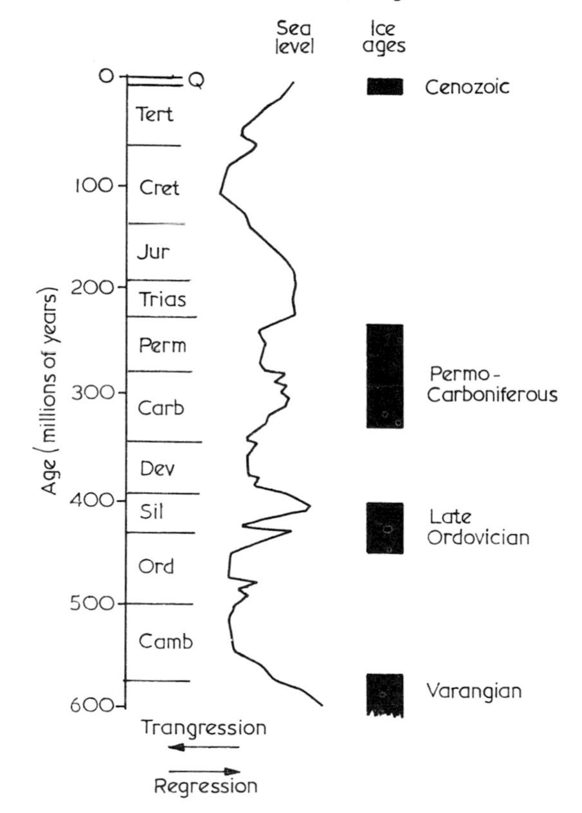

(i.e., times when the 'north' needle of your compass points south) have coincided with major climatic changes such as those responsible for the initiation and ending of ice ages. The magnetic field is determined by the electrical currents generated in the Earth's liquid nickel-iron core, which acts rather like a dynamo. Under 'normal' conditions the magnetic field is directed towards the north; the inclination of magnetism is high in polar areas, parallel with the surface in equatorial areas, and intermediate between the two extremes at intervening latitudes. From the dip of magnetization in rocks of known age, information can be gained concerning their 'palaeolatitudes'.

The occurrence of magnetic reversals is difficult to explain, but they may be due to sudden convection 'plumes' or 'surges' in the Earth's molten interior. These plumes could affect the dynamo so that a sudden change occurs in the direction of electrical current flow—in other words, there is a magnetic reversal.

There have been many changes in the frequency of reversals over the past 2,000 million years. Over the past 80 million years there have been at least 171 reversals, and about 45 million years ago the reversal frequency doubled to its present rate of five reversals per million years. There are some traces of regular oscillations in the reversal frequency, and the geomagnetic changes of the current ice age have often been linked by Earth scientists with glacial-interglacial cycles. But, on a longer timescale, the link between magnetic reversals and ice ages remains to be demonstrated convincingly.

Externally generated causes

1. *Astronomical changes.* Over the years, very many authors have suggested links between the Earth's changing climate and observed or assumed variations in astronomical factors. These variations include changes in solar radiation due to internal solar changes, changes in the geometry of space, changes in the arrangement of the planets with respect to the Sun, galactic changes, and so on.

In most of the astronomical theories of climatic change the Sun is of prime importance. The distinguished astronomer Öpik visualized the Sun as a slightly variable star; that is, that its output of energy is not constant. As the composition of the Sun changes, so there are changes in its luminosity, and it has been suggested that in pre-Quaternary time the solar output of energy was only 85% of its present value. As the Sun burns up its nuclear fuel its surface material is subject to convection, and it undergoes regular alternations of sluggish and dynamic circulation. During sluggish phases there is a slight fall in radiation energy, and during dynamic phases radiation energy increases. There may also be a slow 'flickering'—on a timescale of the order of 250-300 million years—due to the gravitational disturbances of other solar systems as they move through the Galaxy.

Changes in solar emission are widely accepted as having a profound effect upon world climate. Many different periodicities have been recognized or assumed, ranging from millions of years to less than a year. Apart from the long-term oscillation mentioned above, very many medium-term oscillations have been discussed in the specialist literature. Solar emission cycles with periods of between 200,000 and 400,000 years have received particular attention.

On the shorter timescale, climatologists and astronomers have been particularly concerned with the sunspot cycles which have been observed and

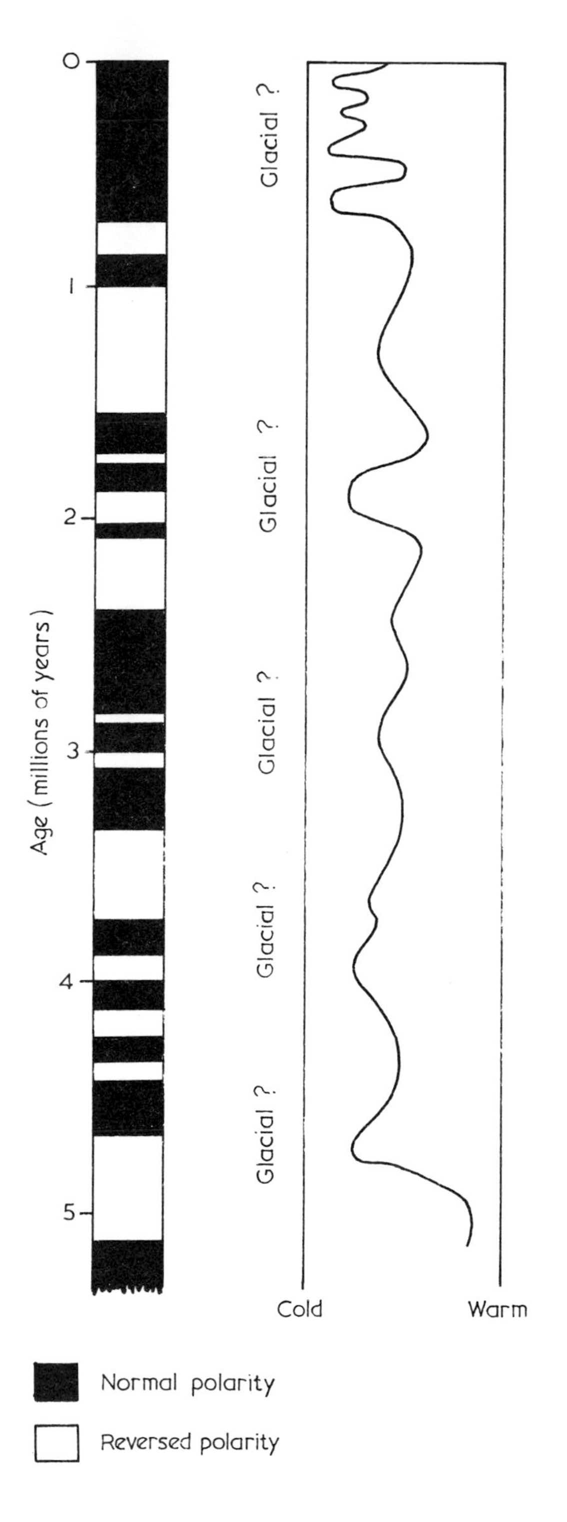

Magnetic reversals during the current ice age. Regular oscillations can be discerned, and these have been linked with glacial-interglacial oscillations of climate.

measured over the last century or so. Cycles with wavelengths of 11, 22 and 44 years seem to be important, but longer-wavelength cycles are difficult to establish scientifically. Also, it has been argued forcefully that the changes in the solar constant (average output of radiation from the Sun) which have preoccupied some climatologists may be more apparent than real. Many 'observed' changes may be quite unreliable, falling well within the range of error of the methods of measurement used.

In spite of the very limited data at our disposal, there have been many computer analyses in recent years of the relation between solar emissions and global climate. Some modelling results show that a reduction in the solar constant of 2-5% would be sufficient to start a new phase of global cooling and icesheet growth, but this prediction can be no more reliable than the data used and the relationships assumed by the person responsible for the computer programme. Indeed, other computer modelling results suggest that a 2-5% reduction in the solar constant would result in *no* icesheet growth.

Another set of ice-age theories concentrates upon changes in the geometrical relations between Earth and Sun. Here the basic principles concern the obliquity of the plane of the ecliptic (that is, the angle between the plane defined by the Earth's orbit around the Sun and the plane defined by the Earth's equator), the precession of the equinoxes (the slowly changing position of the intersection of the plane of the ecliptic with that of the equator), and the eccentricity of the Earth's orbit (that is, the amount by which it departs from the circular).

The most important work in recognizing astronomical periodicities was done by the Yugoslav geophysicist Milankovich in the 1920s. More recently, calculations by others, and especially by Professors A. D. Vernekar and A. Berger, have confirmed the occurrence of the following frequencies:

The precession of the equinoxes: 21,000-25,000 years

The obliquity of the ecliptic: *c.* 41,000 years

The eccentricity of the Earth's orbit: 90,000-100,000 years

Each of these factors might lead independently to small coolings and warmings of the global climate, but substantial climatic changes such as those involved in the initiation of an ice age can probably not occur until all three factors are in 'favourable' phases at the same time. Concerning the combined effects of the three oscillations, it has been calculated that they should be more important in modifying the temperature differences between different latitudes than would be changes in the solar constant. The calculated scale of temperature change seems to match up quite well to the deduced scale of temperature changes for the glacial and interglacial stages of the Quaternary. Milankovich himself took this for granted, and published sets of figures which have been highly influential over the last two decades in particular. The curve we show at the top of page 43 depicts variations in the amount of solar radiation reaching the Earth's atmosphere at latitude 65°N over the past 600,000 years. Curves have also been published by other authors for other latitudes and the diagram shows a calculated curve for 50°N.

In recent years there has been great support for the 'Milankovich theory' from several directions. In particular, a number of studies have been published which show how changing climatic conditions are recorded in deep-sea sediments. In certain marine fossils the amount of 'heavy' oxygen can be related to

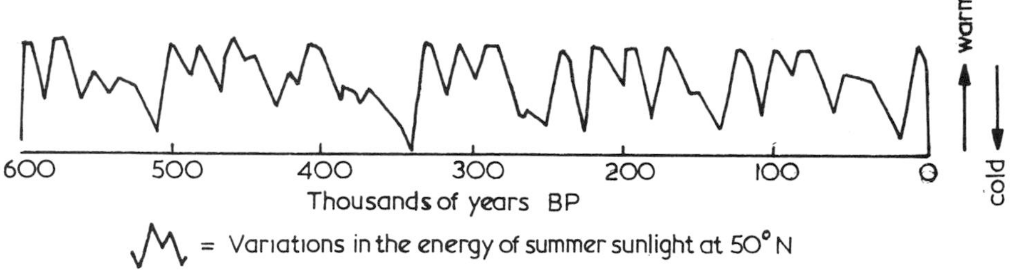

Curve showing variations in solar radiation receipt at the Earth's surface over the past 600,000 years. This is a modified version of one of the curves calculated by Milankovich.

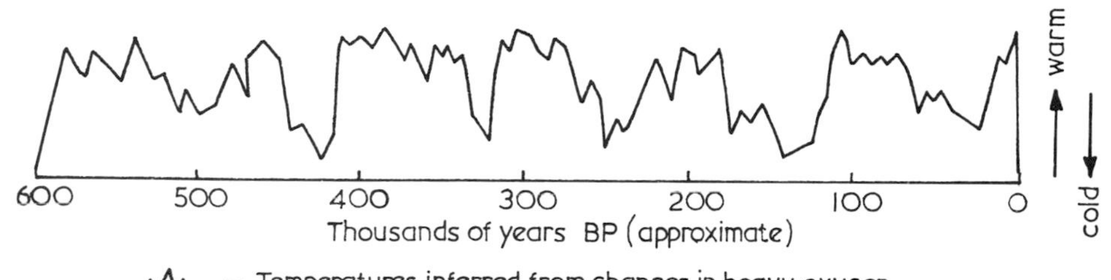

Changes in heavy oxygen concentration in marine organisms (after Hays, Shackleton, Imbrie). Note the approximate correlation with Milankovich 'cycles'.

the amount of ice present in the world, and a plot of changing heavy-oxygen concentrations (and hence warm and cold phases) matches up remarkably well with the plot of warm and cold phases constructed from Milankovich's data (see Chapter 7). This would appear to confirm that changes in the Earth's orbital geometry are fundamentally linked with Quaternary changes of climate.

On the other hand, some scientists are still not convinced, arguing that a simple 'matching up' of graphs is no demonstration of a *real* cause-and-effect relationship. Also, it is claimed that the combined effects of the three astronomical factors considered by Milankovich are likely to be less than the normal seasonal changes which occur on the Earth's surface year after year; if this is so, surely these astronomical effects must be inadequate to explain the onset and ending of ice ages?

On a much longer timescale, galactic changes may be responsible for the apparently regular occurrence of ice ages through geological time. Our Galaxy swirls through space with a period of about 200-250 million years, and this is of the same order of magnitude as the period of the 'geological revolutions' recognized by some geologists in the geological column. As the arms of the Galaxy swirl through space, the position of the Solar System moves up and down through the galactic plane; hence it must be subject to changes in gravitational effects, the effects of interstellar dust veils, and the effects of solar winds from other stars.

There may also be changes in the spin-rate of the Earth, and Professor Rhodes Fairbridge has argued that periods of slowing in the spin-rate may coincide with the occurrence of ice ages. In addition, galactic changes may be responsible, at least in part, for the 'flickering' of the Sun and for intermittent long-term reductions in the emission of solar energy.

2. *Atmospheric changes.* The character of the Earth's atmosphere is delicately adjusted to today's climatic conditions. If this character is changed—for example, through a change in the relative proportions of atmospheric gases—it is highly probable that the climate will change also. A large increase in the amount of water-vapour would affect climate through an increase in the thickness and extent of the cloud layer, leading in turn to a reduction in the amount of solar energy reaching the Earth's surface. On the other hand, a *slight* increase in water-vapour would lead to a *raising* of global temperatures, since the 'heat-retaining capcity' of the atmosphere would be increased. Hardly any atmosphere-climate relationships are simple or direct; the workings of the climate machine are exceedingly complex, being full of indirect or side effects, feedback mechanisms and so forth. Climatologists are only just beginning to unravel some of the complexities of present-day climate, and it is therefore not surprising that statements about past climates are wide open to debate and argument.

Many theories of climatic change stress the importance of carbon dioxide, CO_2, levels in the atmosphere. Most attention is concentrated on the so-called 'greenhouse effect', in a situation where most of the Earth's incoming radiation is short-wave and most of the outgoing radiation is long-wave. CO_2 absorbs long-wave radiation, and heat is retained in the atmosphere rather than being returned into space so long as there is a reasonable amount of carbon dioxide present. (A dramatic example of the effect is shown by the heavily clouded planet Venus, at whose surface rocks glow redly with their own heat, so drastic is the greenhouse effect there.) If the CO^2 content of the atmosphere is reduced, long-wave radiation is not 'trapped' to the same extent, and so temperatures fall.

According to one respectable theory, if the atmospheric CO_2 concentration were to be halved, the Earth would be iced up all over. On the other hand, another equally respectable theory states that a halving of the CO_2 concentration would lead only to a 3C° reduction in global temperatures, illustrating the uncertainties of calculations and predictions in the field of atmospheric science!

If the CO_2 content of the atmosphere (0.03%) were to be increased by forest fires or volcanic activity, the climate would be warmed. This would lead to increased plant growth. Plants would then take up more and more CO_2, leading eventually to a reduction in atmospheric CO_2 levels and thus to a reduction in temperatures. The water-vapour content of the atmosphere would drop, and this would lead to a further temperature reduction. With reduced temperatures there would be a decrease in plant growth, and plants would die and decay in many areas which were previously well vegetated. This would return CO_2 to the atmosphere. In turn this would increase the atmosphere's ability to trap outgoing long-wave radiation, and temperatures would rise gradually again. Thus there may be a long-term oscillation in the amount of CO_2 present in the atmosphere, with regular rises and falls of a set magnitude. A graph of these rises and falls could also show rises and falls of global temperature, lagging slightly behind, and increased and decreased in world rates of plant growth, lagging even further behind. But it will be apparent from the foregoing that cause-and-effect relationships are very difficult to establish; the carbon cycle is notoriously complicated in any case, involving as it does plants, animals, volcanic gases, chemical weathering, carbonate fixation in rocks, and carbon concentrations in the oceans.

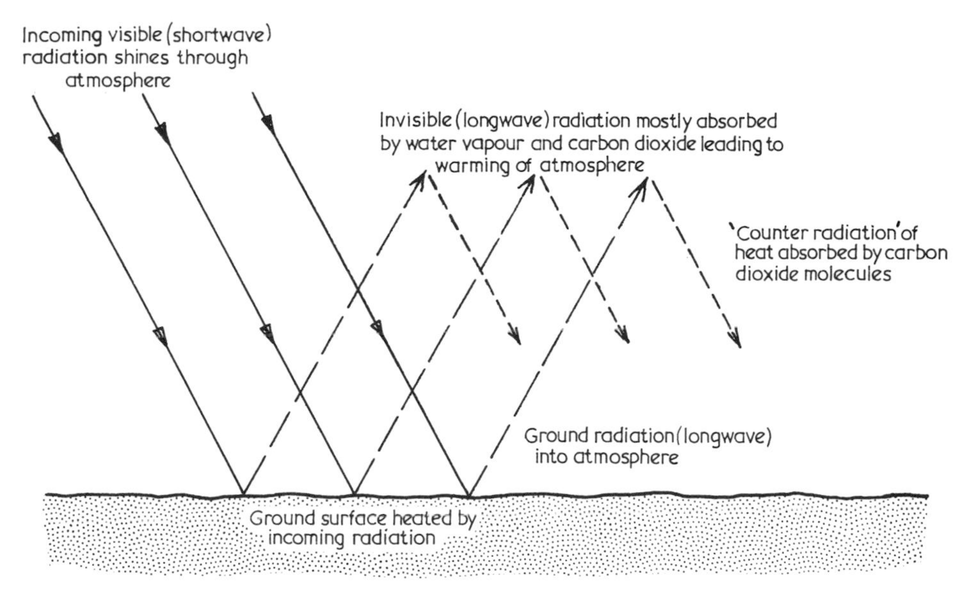

Diagram to illustrate the 'greenhouse effect' by which the Earth's atmosphere is heated with the aid of carbon dioxide molecules.

There has been much comment in recent years about the rise in the atmospheric CO_2 content linked with the burning of fossil fuels. Some climatologists argue that this has caused a global temperature increase of at least 1C°. On the other hand, it is extremely difficult to establish this as a fact. A supposed global temperature rise of 1C° may in reality be a series of regional temperature increases, restricted to certain areas where meteorological stations are thick on the ground and caused simply by shifts in the pattern of atmospheric circulation. Other areas, for example in the southern hemisphere oceanic zone, may be experiencing an equivalent temperature *decrease* which is unrecorded or unrecognized because of the shortage of permanent meteorological recording stations. Even if there is a slight global warming at present, it would be difficult to prove that this is caused by industrial atmospheric pollution. Global climate is never stable in any case, and short-term oscillations of a few centigrade degrees are probably a perfectly normal feature of our environment.

Other atmospheric changes have seriously been considered as factors in the waxing and waning of the world's ice cover. The ozone composition of the upper atmosphere is thought to be affected by solar flares and also by industrial pollution, but it is not certain how much global cooling and warming can be accounted for by changes in the ozone layer. Volcanic eruptions may also affect the climate, especially if there are many large-scale eruptions over a short space of time which add volcanic dust to the atmosphere. However, it is now thought that the effects of volcanic eruptions are short-lived, and it has been demonstrated that volcanic dust seldom remains in suspension in the atmosphere for more than seven years after an eruption.

3. *Geochemical changes*. There are a number of theories concerning the chemistry of the lithosphere, the hydrosphere and the atmosphere which may help to

Large volcanic eruption in Southern Chile in May 1960 adding a column of suspended dust to the atmosphere.

explain the occurrence of ice ages. Like the theories mentioned above, these are closely concerned with the rôle of CO_2 in the atmosphere, but they are also concerned with 'free' oxygen (i.e., oxygen which is not combined with other elements) and with the deposition of carbonates.

It is now known that the oxygen content of the atmosphere has increased irregularly through geological time to its present level of 21%. During early

Precambrian times there were only minute quantities of oxygen in the atmosphere, but by about 600 million years ago the oxygen level had reached 1%, with a consequent evolutionary 'explosion' in the primitive lifeforms of the oceans. At the same time, there was a sharp *decrease* in the amount of atmospheric carbon dioxide following the deposition of large amounts of carbonate material as 'dolostone' (see Chapter 4) in the oceans. Could this change in atmospheric composition have been linked to an increase in atmospheric circulation and an overall cooling of climate? Could the 'new' atmosphere have had anything to do with the occurrence of late Precambrian ice ages?

Later on, during middle and late Silurian time, the oxygen content of the atmosphere reached a second critical level at 10%. Now the lethal effects of ultraviolet radiation were 'shaded out' so that life could exist at or near the ocean surface. By the end of the Silurian period many forms of life had come ashore, and there was a second evolutionary explosion. All this occurred shortly after the end of the Ordovician-Silurian ice age. Could this ice age have been linked with another sharp reduction in the amount of CO_2 in the atmosphere and another phase of global cooling?

During Carboniferous time, according to some authorities, the oxygen content of the atmosphere was greater than it is now, with a corresponding reduction in the amount of carbon dioxide. Much of the abstracted carbon dioxide was precipitated in the oceans as carbonate rock or limestone; much more was locked into the thick Coal Measures which were laid down on the continental peripheries. Again this phase of carbon 'abstraction' can be linked with an ice age—in this case the Permo-Carboniferous ice age of 300 million years.

Finally, the cooling of the Tertiary period, which has culminated in the Quaternary ice age, seems to have followed another phase of carbonate deposition on a massive scale which formed the widespread Cretaceous chalk deposits and the Tertiary corals and deep-sea oozes. Again the links in the chain reaction might be as follows: carbonate accumulation in the oceans; reduction of atmospheric CO_2; lowering of world temperatures; onset of an ice age.

4. *Oceanic changes.* The ocean exerts a very strong control over global climate, acting largely as a huge reservoir of heat. Ocean currents also serve to transfer considerable quantities of heat from tropical to polar lands, while cold currents moving away from the high latitudes have a cooling effect upon the landmasses with which they come into contact.

During Oligocene times (38-26 million years ago) Australia became separated from Antarctica while the land link between South America and the Antarctic Peninsula was also broken. These events led to a great change in the oceanic circulation of the southern hemisphere; for the first time ocean currents were able to circulate freely around the Antarctic continent, linking the waters of the Pacific, Indian and South Atlantic Oceans in the process. The development of circumpolar ocean currents and atmospheric circulation reduced the import of warmth from the equatorial and middle latitudes, and during Miocene times the Antarctic icesheet grew until it was considerably larger than at present (see Chapter 7). The Tertiary period saw also the gradual widening of the North Atlantic, an event which had a great effect on oceanic circulations and global climate. Until about 3½ million years ago there was no land link between North

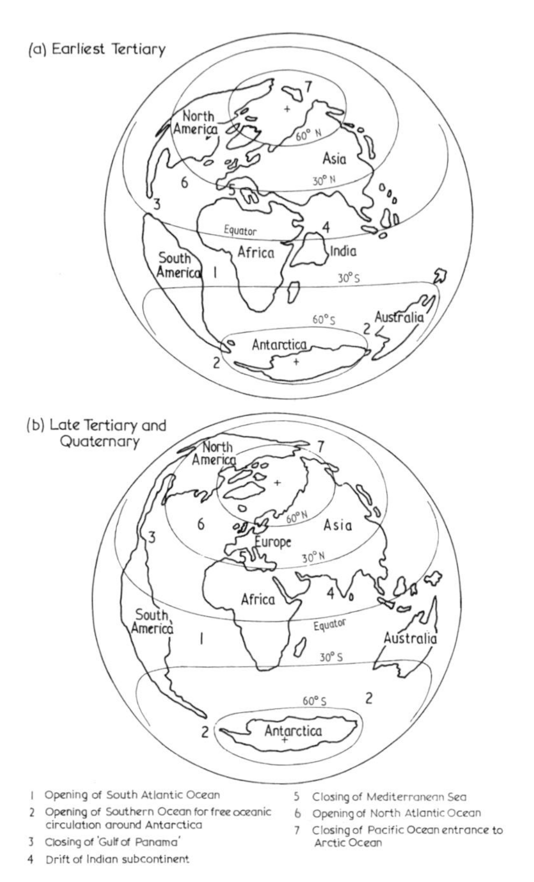

Some of the palaeogeographic changes of the Tertiary period which influenced oceanic circulation.

America and South America, and the raising of the Isthmus of Panama may have caused great changes in the pattern of oceanic circulation in equatorial latitudes. During Miocene and Pliocene times the link between the Mediterranean Sea and the Indian Ocean was cut off, and the Straits of Gibraltar were opened. In addition, the islands of the East Indies began to emerge from the sea, greatly reducing the exchange of water between the Pacific and Indian Oceans.

The present-day pattern of ocean currents became established as a result of the palaeogeographical changes mentioned above. Around Antarctica a vast reservoir of cold 'bottom water' was produced, and this was not the only ocean area where temperatures were reduced. The virtual closing of the old gulf between the Arctic and Pacific Oceans reduced the import of warm water from the equatorial zone and also made it difficult for cold water to escape. In addition, the northward flow of many large rivers from Canada and Siberia introduced great amounts of cold fresh water, reducing the salinity of Arctic waters. The net result was a rapid cooling of the ocean and the creation of a cover of pack ice. The North Atlantic Drift, which previously carried most of its warm waters into the Labrador Sea, was largely diverted eastwards. Greenland ceased to be warmed by this current, and it was further cooled by the development of a cold current flowing southwards along its east coast from the Arctic Ocean. About 3.4

million years ago the Greenland icesheet developed to something like its present size.

The present configuration of landmasses and arrangement of ocean currents could be altered substantially by changes in sea-level or by the uplift of sections of the sea floor. One ice-age theory, proposed by Ewing and Donn in 1956, used the idea that a slight world-wide fall of sea-level, caused by a combination of glaciation and other factors, could cause such a shallowing of water over the Faroe-Icelandic sill (a ridge of basalt stretching from Iceland across the North Atlantic towards the Norwegian coast) that the import of warm water into the Arctic would be greatly reduced. This in turn would promote cooling and ice extension in the Arctic Ocean once a glacial period was under way. It is known that, at the peak of a glacial period, sea-level falls by 100m or more, and inevitably the emergence of 'land bridges' in other areas also would have a great effect upon the pattern of oceanic circulation.

However, in this field as in others, the recognition of cause-and-effect relationships is extremely difficult. It has been pointed out that these latter oceanic influences upon the course of an ice age cannot operate until glaciation has already been responsible for the abstraction of a great deal of water from the oceans. They should therefore be looked on as effects of glaciation, rather than causes. On the other hand, longer-term changes in the circulation of the oceans do seem to have been closely related to the onset of the Quaternary ice age, and similar changes may have helped to trigger off earlier ice ages also.

There have been a number of ice-age theories which use the idea that interglacial oceanic conditions may in themselves be responsible for the initiation of glaciation. One such theory uses the idea of an ice-free Arctic Ocean as a starting-point. With open water in the Arctic, abundant moisture could be picked up by the polar winds to be precipitated as snow on the mountains and plateaux around the fringes of the ocean. Glaciers, ice caps and icesheets would begin to expand, and the resultant climatic cooling would lead eventually to the freezing over of the Arctic Ocean. This would be helped, as the northern-hemisphere icesheets grew to full size, by a lowering of sea-level and a reduction in the amount of tropical water flowing into the Arctic basin. Eventually, when the Arctic Ocean was completely covered by icesheets, ice shelves and pack ice, there would be no oceanic moisture available for the nourishment of the glaciers in the north polar regions, and the glaciers would waste away rapidly as a result. Hence both the onset and the ending of a large-scale glaciation can be explained by reference to 'oceanic' factors.

The Behaviour of Ice

In recent years a number of glaciologists and geomorphologists have been making the point that the behaviour characteristics of ice (which is, after all, the basic raw material of which ice ages are made) may in themselves be responsible for the intermittent expansion and contraction of the world's icesheets. By implication, the oscillations of climate associated with ice ages may be not so much the causes of glaciations as the effects. In other words, an expansion of the world's ice cover *causes* a cooling of climate, and a reduction in the ice cover *causes* a global climate warming. For many people these are difficult ideas to accept, for in the conventional wisdom of ice ages it is almost always assumed that mechanisms of climatic change are needed before the occurrence of ice ages can be explained.

An understanding of a few simple glaciological facts may help to diminish our faith in 'climatological' ice-age theories. In the first place, as mentioned on page 34, glaciers do not necessarily develop most easily in polar locations. The most favourable locations for glacier growth are not the cold dry polar regions but the cool moist middle latitudes, where there is abundant precipitation and where conditions are cold enough during the summer to allow at least some of the winter accumulation of snowfall to survive and build up year by year. In the second place, large present-day glaciers (such as the Greenland and Antarctic icesheets) are capable of exerting a strong control over their own climates, and there is no reason to assume that ancient glaciers would have been any different. A large expanse of snow and ice with a very high surface whiteness or albedo reflects much incoming radiation back into space, in contrast to a dark surface which tends to absorb heat. Hence icesheets have a strong cooling effect upon the atmosphere. Some of this cold air drains off the ice surface, funnelled along outlet glaciers and other depressions as the ferocious katabatic winds which polar explorers have always feared. These winds tend to have a cooling effect beyond the edges of icesheets and ice caps, thus increasing the size of the 'cooled area'. In addition, a large ice mass, once established, tends to 'block' incoming cyclonic storms and direct them into lower latitudes. In the middle latitudes of the northern hemisphere this leads to an increase in storminess, greater precipitation and also possibly a lowering of mean annual air temperatures. In turn this creates a favourable environment for the growth of mid-latitude glaciers.

All of the above implies a gradual shift from a 'non-glacial mode' to a 'glacial mode', where the cause-and-effect relations are largely to do with ice and its influence upon the atmosphere.

Conventionally, the onset of glaciations or ice ages has been considered to be a slow process, with thousands of years required for the build-up of icesheets once they have been initiated. However, some modern theories refer to an almost instantaneous 'flip' from non-glacial to glacial conditions (see Chapter 7). One such theory has been popularly labelled the 'snowblitz' theory. According to Nigel Calder the mid-latitudes are vulnerable to a sudden switch of climate from relative warmth to harsh and increasing cold. He proposes that the green meadows of England could easily be affected by a run of severe, snowy winters. If the snow-cover (or part of it) were deep enough and extensive enough to survive during a run of cool summers, then the increased albedo of the land surface could lead to a sharp lowering of air temperatures. As more precipitation falls as snow, the snow-cover increases in thickness, and summers become so cool and so short that ablation rates are incapable of keeping pace with the rate of snow build-up. Once this point is reached there is nothing to stop the thickening of snowbanks until glacier ice is created. Thus glaciation can commence unexpectedly, over the space of a few years, with the glaciers developing not in the remote snowy highlands but on the lowlands and plateaux of the middle latitudes (55-65°N).

While this idea undoubtedly has a certain popular appeal, it has been greeted with scepticism by most glaciologists and climatologists, who believe that the energy balance of the middle latitudes could not be disturbed to the extent

White snow and dark rock—part of the Shackleton Range projecting through the surface of the Antarctic icesheet. Snow has a high surface albedo which reflects much solar radiation back into space.

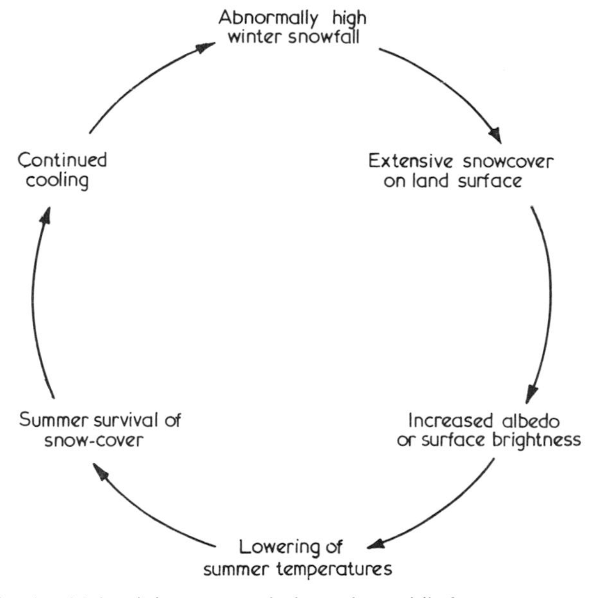

The positive feedback which might operate during a 'snowblitz'.

envisaged by Calder. On the other hand, the idea of *high-latitude* snowblitz is much more likely, with snowy winters and cool summers leading to a rapid snow build-up in Arctic Canada or Lapland. During the early 1970s, satellite monitoring of the ground surface of Arctic Canada showed that some parts of the far north were becoming covered by perennial snow, whereas previously the snow-cover had melted away each summer. The snow build-up did not reach 'take-off point' on that occasion, and a slight warming occurred during the mid-1970s to reduce the extent of the snow-cover again.

However, sudden climatic coolings are known from the records of past climates. One such occurred about 89,500 years ago during the last interglacial, when global temperatures dropped dramatically during no more than a century or two until they were as low as those of a glacial stage. After a few hundred years the temperatures rose again. There is now much speculation about the nature and cause of this climatic event, but it was certainly sudden and it probably brought the world to the brink of full-scale glaciation. There have been other sudden, but less severe, coolings within the past 13,000 years, and we cannot afford to reject the snowblitz theory out of hand in our efforts to explain them.

One influential ice-age theory which concerns the behaviour of ice is the 'surge theory' of A. T. Wilson. In 1964 he proposed that periodic surging of parts of the Antarctic icesheet could occur, transporting vast amounts of ice from the icesheet interior to the coast and leading to the creation of a great ice shelf encircling the continent. This ice shelf would displace sea water, leading to a sudden rise of sea-level. It would also lead to a great increase in the ice-covered area of the Southern Ocean, pushing the Antarctic pack-ice fringes much further north than those shown in the satellite photograph on page 14. Furthermore, there would be a disruption of circumpolar ocean currents. The albedo of the southern hemisphere would rise sharply, leading to hemispheric and then global cooling. In turn, this global cooling would cause an increase in the ice cover of the northern hemisphere, culminating in the growth of the Laurentide, Scandinavian and other mid-latitude icesheets. Sea-level would fall gradually to its 'glacial' low position, thus helping to stabilize the Antarctic ice shelf on previously submerged reefs and shoals.

On page 34 it was noted that glaciers can grow as a consequence of climatic *warming*. This apparent paradox has to do with the fact that for their growth glaciers require abundant moisture rather more than low temperatures, so that moist cool conditions are more favourable for the inception of an icesheet than are cold dry conditions. Warm or interglacial conditions may also be required in order to trigger an Antarctic icesheet surge.

In order to understand this statement one needs to appreciate how ice temperatures are distributed in a large glacier or icesheet. In general, ice temperatures rise with depth beneath the surface. If the surface of an icesheet is very cold (say −30°C) then ice temperatures will be about −5°C at a depth of 2,000m. The ice will still be cold enough in this case to remain frozen to the bed. However, if the ice is thicker or if the surface temperatures rise, the temperature gradient in the ice will eventually be altered so that the ice temperatures at the base rise above a critical threshold. This threshold is not necessarily 0°C, but the 'pressure melting point' of the ice, which may be as low as −1.5°C. Once

The snout of a surging glacier in west Greenland. This illustrates, on a small scale, the conditions that would prevail if part of the Antarctic icesheet were to surge.

the pressure melting point is reached, basal melting of the icesheet can occur, and the ice begins to slide on its rock bed. The previously stable icesheet now shows signs of instability, and there is a likelihood of surging behaviour in which sections of icesheet move at many times their normal velocity, transporting huge quantities of ice to the icesheet edge and leading to a subsidence of the icesheet surface in the area where the surge commenced. When the ice has thinned sufficiently, basal ice temperatures fall beneath the pressure melting point again. The ice freezes onto its bed again, and the icesheet reverts to a stable condition. After this, the icesheet surface slowly begins to build up again until, after many thousands of years, it acquires the thickness and temperature characteristics required for the next phase of surging.

The above sequence of events is quite plausible from a glaciological point of view, and some glaciologists now believe that the Antarctic icesheet, or parts of it, behaves in a cyclic fashion as shown below. This cycle of behaviour is largely independent of climate and climatic change, and it may provide us with a perfectly plausible *glaciological* mechanism for the glacial and interglacial periods which occur within a normal ice age.

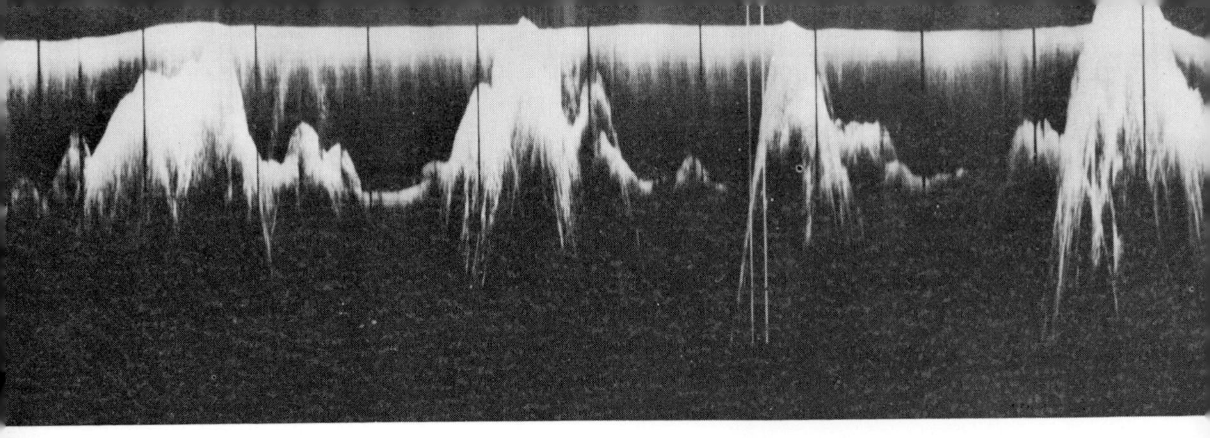

A radio echo sounding profile through part of the Antarctic icesheet, showing buried mountains and valleys and the icesheet surface. Where the ice is thickest, in the deep troughs, it may be quite close to its 'pressure-melting point'.

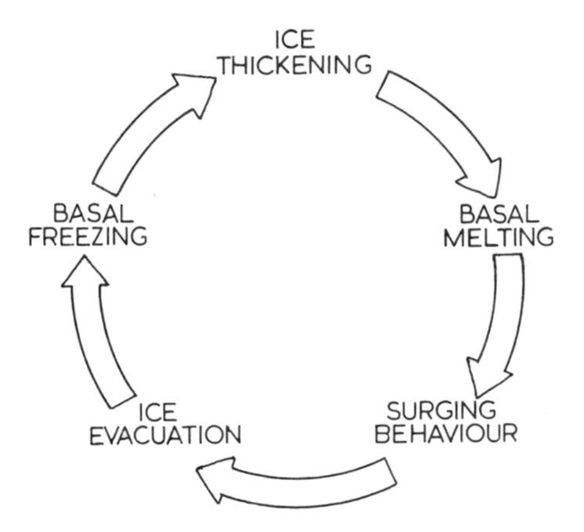

The 'surge cycle' which might affect parts of the Antarctic icesheet.

There are many difficulties associated with the icesheet surge theory, not least of which is the problem of the great Antarctic ice shelf which is supposed to be created following an icesheet surge. How is the ice shelf stabilized on an open coastline, and how is it broken up at the end of a glacial period? These are questions which remain to be answered, and as in the earlier sections of this chapter it is extremely difficult to differentiate between causes and effects. For example, is the break-up of the ice shelf a cause of climatic warming, or is it a consequence of a temperature increase and a rise in sea-level resulting from the breakdown of the northern-hemisphere icesheets?

Whatever the answers to such questions may turn out to be, evidence is now accumulating in support of certain parts of Wilson's surge theory. It is known that large areas around the peripheries of Antarctica have basal ice temperatures at the pressure melting point, and radio echo-sounding investigations have revealed that parts of the West Antarctic icesheet fringing the Ross and Filchner ice shelves are underlain by large subglacial lakes. The bedrock in several such areas lies well beneath sea-level, and it would be relatively easy for the icesheet to surge and perhaps even disintegrate if the basal water layer became too extensive or if the sea penetrated far enough inland to create ice shelves in areas currently

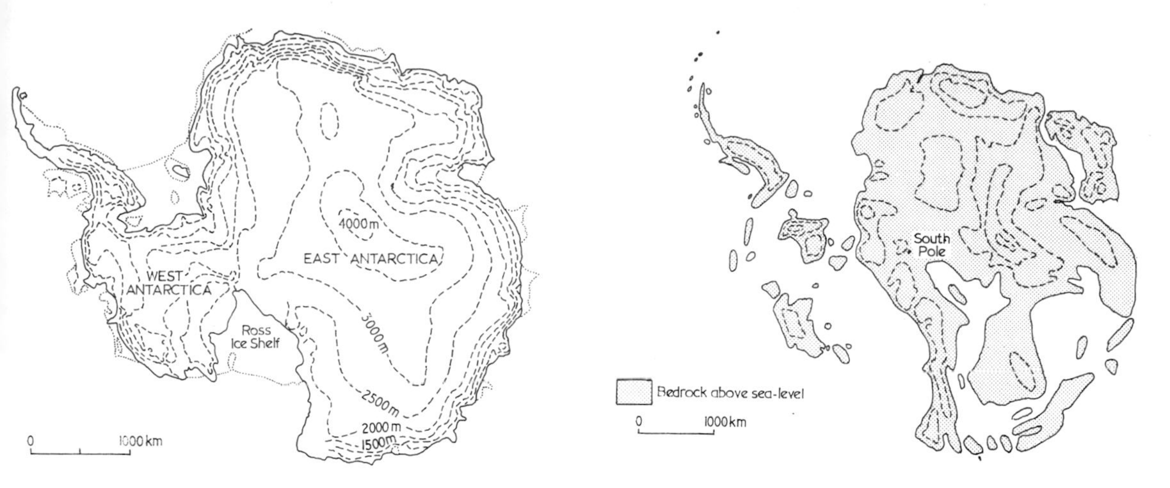

The shape of the icesheet surface (*left*) and the bedrock base (*right*) in Antarctica. Much of the western part of the icesheet is resting on bedrock below sea-level, and this is an ideal situation for 'surging behaviour'.

occupied by the icesheet. The 'slumped' cross-profile of part of the West Antarctic icesheet is already interpreted by some authorities as indicative of surging behaviour, and it is felt that the surface is now building up again in readiness for the next surge. The disintegration of the Laurentide icesheet in North America may happen in a similar way, and John Andrews discusses this in more detail in Chapter 7, citing evidence from 8,000-7,000 years ago for the area around Hudson Bay. Here again, as emphasized on page 49, the relationship between glaciological factors and world sea-level is of critical importance.

Chain Reactions and Feedback Loops

It will have been apparent from the earlier sections of this chapter that there are extremely complicated links in the environmental system composed of the atmosphere, the hydrosphere and the cryosphere. The air, the oceans and the great icesheets are to a large extent related, and a change in the character of one of them is quite likely to lead to changes in the other two. Hence we have discussed the effect of a global temperature reduction in cooling the oceans and leading to an expansion of the icesheets; the effect of a change in ocean currents or a change in sea-level in altering precipitation patterns and causing the growth of glaciers in certain areas; and the effect of an icesheet surge in cooling the atmosphere and leading to a sudden rise of sea-level.

At the moment the environment is experiencing a multitude of chain reactions of different types, different magnitudes and different durations. The single greatest problem confronting Earth scientists and atmospheric scientists is the disentangling of one chain reaction from another and the demonstration of cause-and-effect relationships which have meaning on a long timescale. The basic defect in any reconstruction of environmental change on a geological timescale is that *we have direct measurements of the environment for only a few centuries.* For some critical factors, such as changing ocean temperatures or changing glacier mass balances, we have direct measurements spanning only a few decades. Thus even the most sophisticated theory of ice ages has to depend upon assumptions and reconstructions, usually from periods well before the evolution of mankind. Even the most sophisticated mathematical model of ice ages has to depend upon an input of data of uncertain quality; this is inevitable, for 'data' even for the period around 18,000 years ago (a period studied in detail during the sophisticated CLIMAP project) has to be invented (see Chapter 7). And the invented data can be no better than the sources of information used by the inventor.

On the other hand, cause-and-effect relationships are much easier to discern on a short timescale, and meteorological studies have revealed a number of plausible mechanisms for promoting the onset of glaciation or bringing a glacial period to an end. Two basic assumptions have to be made: that even on a short timescale climatic changes can affect both oceans and glaciers; and that the effects of quite small climatic changes can be 'magnified', having far-reaching effects particularly on ice masses in a critical condition.

The link between weather patterns and glacial expansions and contractions has been studied by many meteorologists and climatologists during the present century, notably Professors Ahlmann, Brooks, Manley, Lamb and Flohn. In recent decades numerical modelling has been undertaken, with computer analysis of meteorological data used as a basis for projections back in time. Three of the projects of greatest importance go by the delightful names GARP[1], NORPAX[2], and CLIMAP, and they are related to patterns of global or hemispheric atmospheric circulation. During the last few years many interesting and influential papers have been written on the theme of numerically based climatic reconstructions of atmosphere and oceanic circulation patterns and their relations with the global ice-cover 18,000 years ago. Some of the research results are now being published; the 18,000-year reconstruction of the pattern of glaciation is discussed later in this book by John Andrews.

A typical meteorological explanation for the onset of an ice age might run as follows. The atmospheric circulation of the northern hemisphere wanders into a persistent 'block' in which cold air is maintained over the lands of the North Atlantic. Depressions are forced to follow more southerly tracks, and cold air drawn from the Arctic increases the snowfall over the middle latitudes. Glaciers and then icesheets begin to grow in Lapland and parts of Arctic Canada. The oceans are cooled, and pack ice is formed at sea well to the south of the Arctic Ocean, notably around the coasts of the North Atlantic and North Pacific oceans. In the Atlantic, oceanic circulation is altered so that the warm North Atlantic Drift is weakened. As the northern-hemisphere ice masses grow, the new pattern of atmospheric circulation is reinforced; moreover, the now persistent oceanic thermal anomaly pattern also reinforces the atmospheric circulation state that produced it. The icesheets grow to full size, and the world finds itself in the grip of an ice age. This sequence of events is referred to as positive feedback, where each link in the chain reaction reinforces the new environmental situation.

The great difficulty with this type of explanation is in finding a mechanism for *ending* an ice age once it has become fully established. Clearly the positive

[1] Global Atmospheric Research Program.
[2] North Pacific Experiment.

feedback has to be stopped somehow if the whole globe is not to become iced up. Some scientists now believe that once the 'glacial' atmospheric mode has become firmly established it loses its stability. Glaciological factors may be partly responsible for this, especially in the case of unstable 'marine' icesheets. Quite simply, once the northern-hemisphere icesheets have reached a certain size, with their southern margins well into the middle latitudes, they become too large to survive. Their interiors are deprived of the snowfall needed to maintain them, and most precipitation is concentrated along the southern and western margins where melting is also greatest. The northern and eastern margins of the icesheets receive hardly any precipitation; the Arctic Ocean and the northern parts of the Atlantic and Pacific Oceans are covered by pack ice, so that the winds which traverse these areas cannot pick up the moisture needed for conversion into snow. Under these circumstances the continued survival of the mid-latitude icesheets is impossible, and catastrophic melting sets in. The polar icesheets of Greenland and Antarctica are largely immune; they survive while the Laurentide, Scandinavian and Siberian icesheets melt away.

Theories such as these are based very much on latitudinal or longitudinal changes in the atmospheric and oceanic circulations of the middle latitudes, and on their relations with the mid-latitude icesheets. The Arctic Ocean, the Antarctic icesheet and the Greenland icesheet play subsidiary rôles, whereas in other theories they assume great importance. Clearly there is as yet no agreement as to where the key to large-scale climatic change really lies—it could lie in the south polar region, the north polar region, or the northern-hemisphere middle latitudes.

Over twenty years ago Professor J. K. Charlesworth was moved to write, when considering the multitude of theories on the causes of ice ages, that they ranged 'from the remotely possible to the mutually contradictory and the palpably inadequate'. Things have become even more confusing since then, although there has been a strong move away from 'catastrophist' theories to theories involving normal environmental oscillations and feedback relations. The meteorological theories expressed in numerical terms and portrayed on computer map reconstructions are typical of these. On the other hand, some modern theories concerning almost instantaneous 'flips' from glacial to interglacial conditions (and back again) smack distinctly of the catastrophic! In spite of the speculation and controversy, the most plausible theories are the 'composite' ones, based upon changes of landmass altitude, continental arrangements, and changes in the amount of solar energy received at the Earth's surface. In the chapters which follow several of these theories are examined again in so far as they are applicable to specific ice ages. At the end of the book some space is devoted to the most difficult question of all—namely the problem of accounting for the apparently regular recurrence of ice ages through geological time, with apparently regular glacial and interglacial periods punctuating each ice age. The periodicity of global warmings and coolings may, when all is said and done, be the fact which allows the real causes of ice ages finally to be isolated and understood.

BRIAN JOHN

3 Glaciers and Environment

To most people, the vision conjured up by the phrase 'ice age' is one of unmitigated cold, with a sparsely vegetated land surface frozen hard, usually snow covered and with massive icesheets and glaciers filling plains and valleys. The picture may include savage and scantily clad primitive men hunting woolly-skinned animals of nightmare proportions, Man himself surviving in dark and dank cave dwellings with the aid of fire, his one great comfort.

While such a montage is broadly based on scientific fact, it is misleading because it emphasizes only the extreme conditions of glaciation in the interiors of the northern continents of the Earth and because it fails to recognize the

Two impressive outlet glaciers carrying ice from the Greenland icesheet towards the coast.

great variety of environments known to have occurred during the Earth's ice ages. It is also misleading because there were a number of ice ages long before the appearance of mammals (including man) on Earth.

What, then, is an ice age? After all, both Antarctica and Greenland are substantially covered by icesheets at the present time, and large areas of many of the world's mountain ranges (particularly those raised during the Alpine orogeny, starting about 65 million years ago) are covered in ice and snow, some of the glaciers and snow fields being of impressive proportions. As indicated in Chapter 1, there is no doubt that we are living in an ice age. At the same time, recognition of landforms and sediments of glacial origin in places like Warsaw, Berlin, Oxford, Galway, Saint Louis, New York and Kiev clearly indicates that the present time is not one of glacial extremes.

The concept of glacial and interglacial stages was introduced in the previous chapter. During each ice age the main glaciations, or glacial stages, are punctuated by periods of milder conditions when glaciers and icesheets diminish. Sometimes the continental icesheets, although reduced in volume, persist and then readvance. The 'stillstands' of the ice margin are referred to as 'stadials', and the warm and relatively short-lived intervals which coincide with glacier retreat are referred to as interstadial periods or 'interstades'. When continental icesheets waste away completely, however, and conditions become temperate in the former glacial zone for many thousands of years, we speak of an interglacial period.

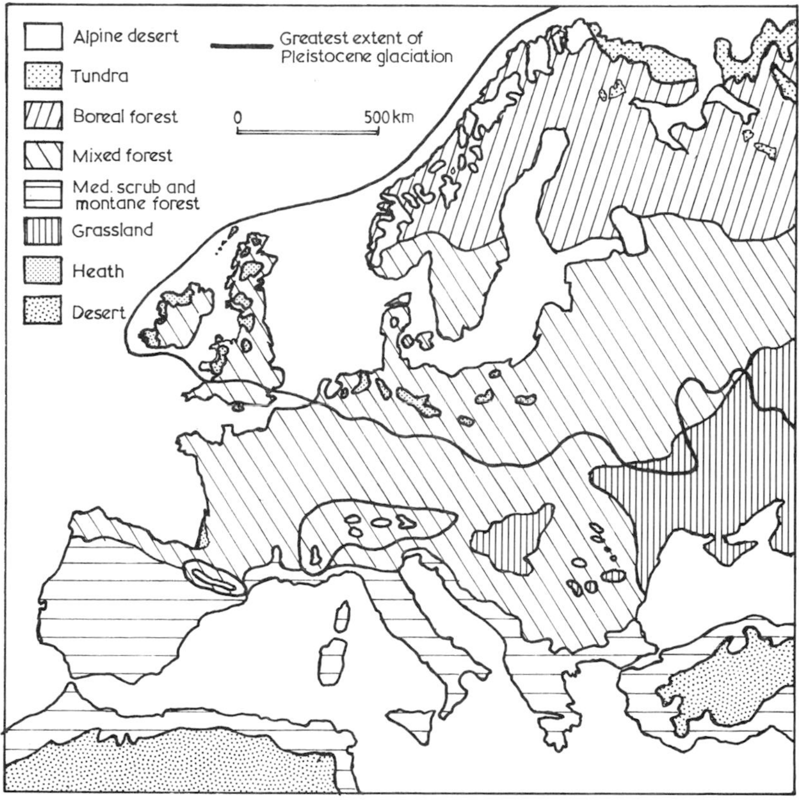

Map of the present distribution of major forests and grassland areas in western Europe, showing their relationships with the maximum ice limits of the Pleistocene epoch.

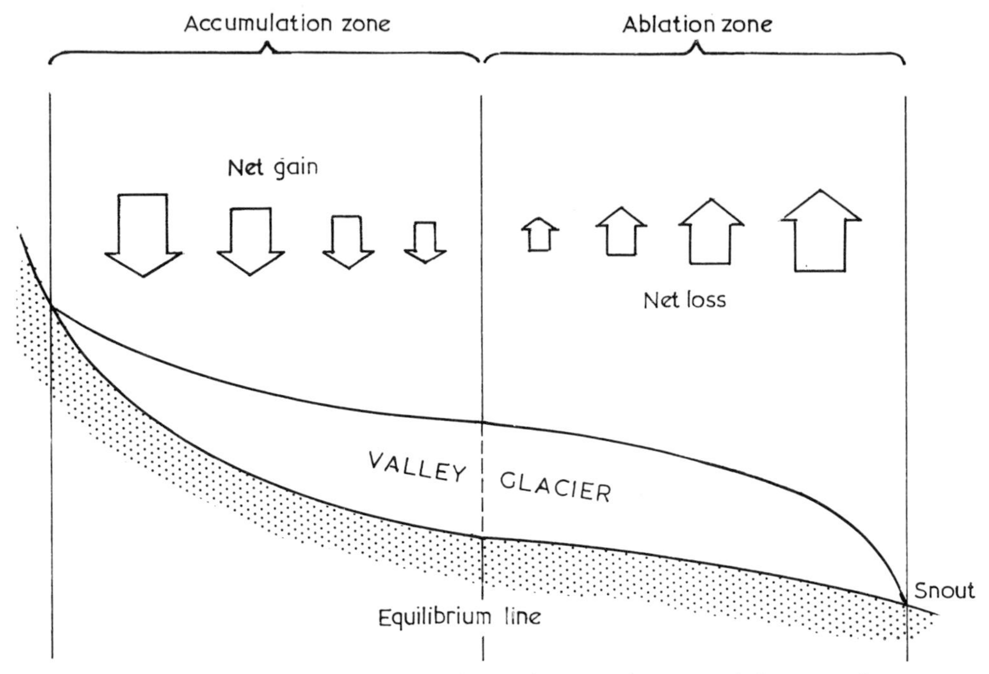

Diagram showing, in simplified form, the relations between the accumulation zone, the ablation zone, and the equilibrium line on a small valley glacier.

We are now in a position to answer the question with which we began. An ice age is a period during which icesheets occupy the plains of the middle latitudes of the Earth, at least from time to time. The waxing and waning of icesheets, however, is an essential property of ice ages, which have a four-fold structure made up of glaciations, interstades, stadials, and interglacials. The limited amount of ice cover on the Earth's surface at present, together with the presence of rich forests and grasslands over much of the ground covered by ice during the last glaciation, are conditions typical of interglacial stages.

Ice on the Land

Our major clues to the nature of the ice ages of the past, and to the environments on Planet Earth during those 'winters', are provided by the effects that exetnsive glaciation has on the Earth's surface beneath the ice, effects made manifest by landforms which remain after the glaciers have gone. We shall, therefore, devote much of this chapter to an examination of features of the landscape which betray the presence, nature and degree of glaciation at some past time, and of the processes responsible for those landforms. Let us look first at glaciers themselves.

To a large extent, glaciers are the products of climate. Glaciers are made largely of crystalline glacier ice, which is created when surface snow is buried more and more deeply by successive snowfalls and is transformed by the pressure of the overlying layers. But no glacier can grow unless, over a period of years, the input of snow (or accumulation) is greater than the combined losses due to melting, evaporation and sublimation (the processes of ablation). In other words, the 'budget' for the glacier has to be positive.

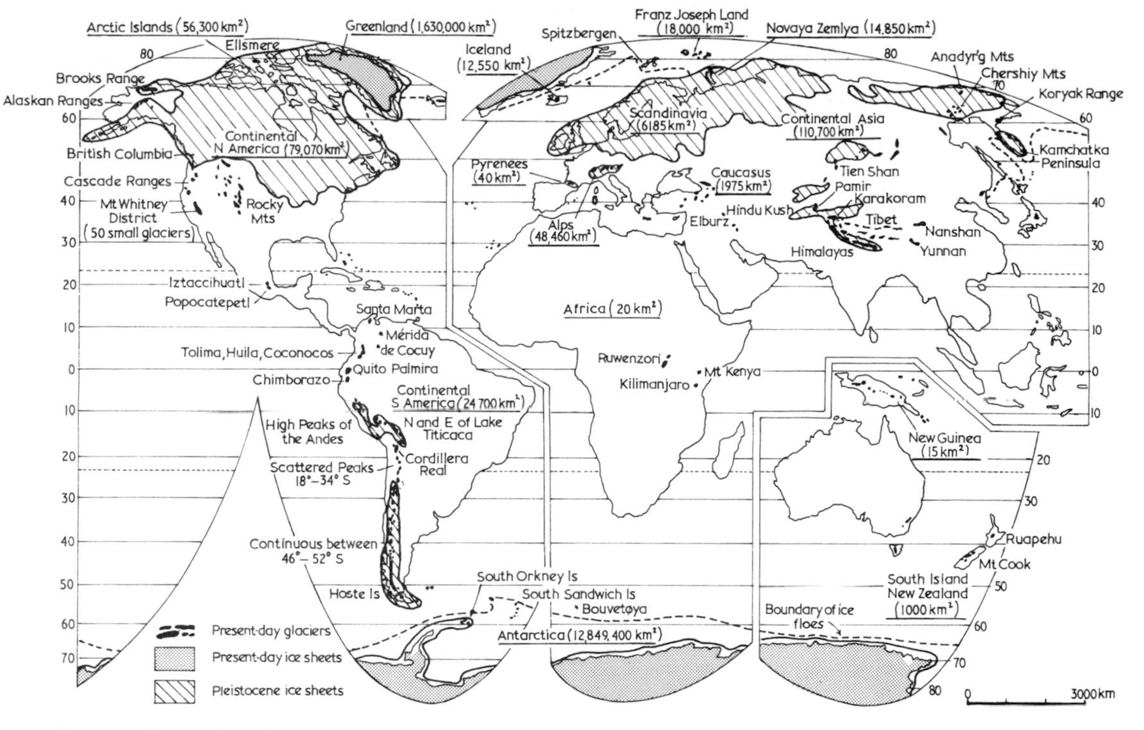

Present-day active glaciers on Planet Earth, and the Pleistocene extent of icesheets.

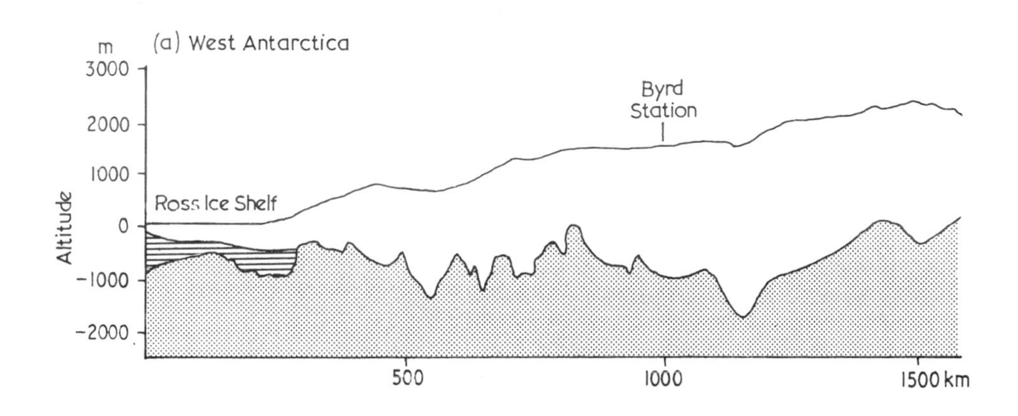

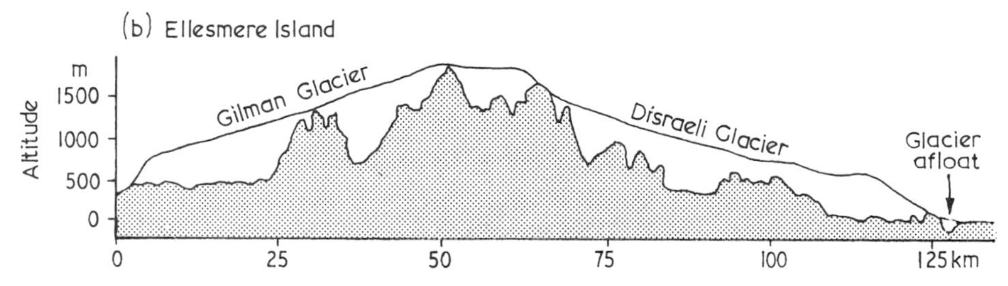

Mountains buried by ice. These profiles from West Antarctica and Ellesmere Island show how icesheets and ice caps are capable of burying *all* of the features of the land surface.

The upper parts of most glaciers are positive in this sense but the lower, marginal zones are often negative, summer losses of volume being greater than the input of snow in winter. In such a case, the upper parts of the glacier are referred to as the accumulation zone and the lower reaches as the ablation zone. The narrow belt separating these two zones is referred to as the equilibrium line because the budget is balanced there. The relative sizes of the accumulation and ablation zones are an indication of the total *mass balance* of the glacier. Changes in the total mass balance resulting from climatic variations are reflected in changes in the altitude of the equilibrium line. In polar regions, the whole of a particular glacier may be above the equilibrium line, whereas in other climates the equilibrium line may be very high, near the head of the glacier. Glaciers which persist for a prolonged period entirely below the equilibrium line become stagnant or 'dead', and they ultimately disappear.

In the early stages of a glaciation, we should expect to find glaciers located in the high mountains of a continent near to the oceans (their source of moisture). Some confirmation of this is provided by the present distribution of glaciers in the world, but, when we look at a map of this, we note that there are exceptions. For example, the great icesheets of Antarctica and Greenland have engulfed mountainous areas entirely, and even quite small icesheets have managed to override at least the lower parts of mountain ranges. It appears that, if an ice mass is large enough to be called an icesheet, it is partly independent of the local geography. It tends to act as a great dome with movement outward from the higher to the lower parts modifying the landscape beneath and, indeed, acting as a very important regional climatic influence in its own right.

Glaciers occur in many sizes and shapes. Small round or oval glaciers occupy high rock alcoves called cirques. When such glaciers expand they overspill their cirques and extend to lower altitudes as valley glaciers. Dome-like glaciers on plateau surfaces are called ice caps, but when they extend to cover a whole region such that their surface form is not dictated by the form of the ground beneath them we describe them as icesheets. Many valley glaciers and the outlet tongues from ice caps and icesheets terminate in lakes or in the sea. The large outlet glaciers of the Ross Sea and other embayments of Antarctica flow together into large floating glaciers called ice shelves.

How Glaciers Work

Temperature plays an important part not only in the establishment and growth of glaciers but also in the way in which they move and the degree to which water melts from them. If glacier ice rests on a slope, its interior will deform over time; the rate at which this occurs varies with the temperature of the ice. Glaciers also move by sliding on their beds; for this to occur, the temperature of the ice/rock contact must be at the melting point, such that a thin film of meltwater is produced. Glaciers in the mountains of the middle latitudes tend to be in this condition; their movement is made up of a combination of deformation and sliding, and they are known as temperate, or warm-based, glaciers.

In polar regions, however, air temperatures may be so low that the temperature of the ice throughout a glacier is well below the freezing point. Such cold-based

(*Opposite*) A valley glacier in east Greenland, fed by ice from a number of cirques in the mountains and flowing towards the lowlands in the trough which it has eroded.

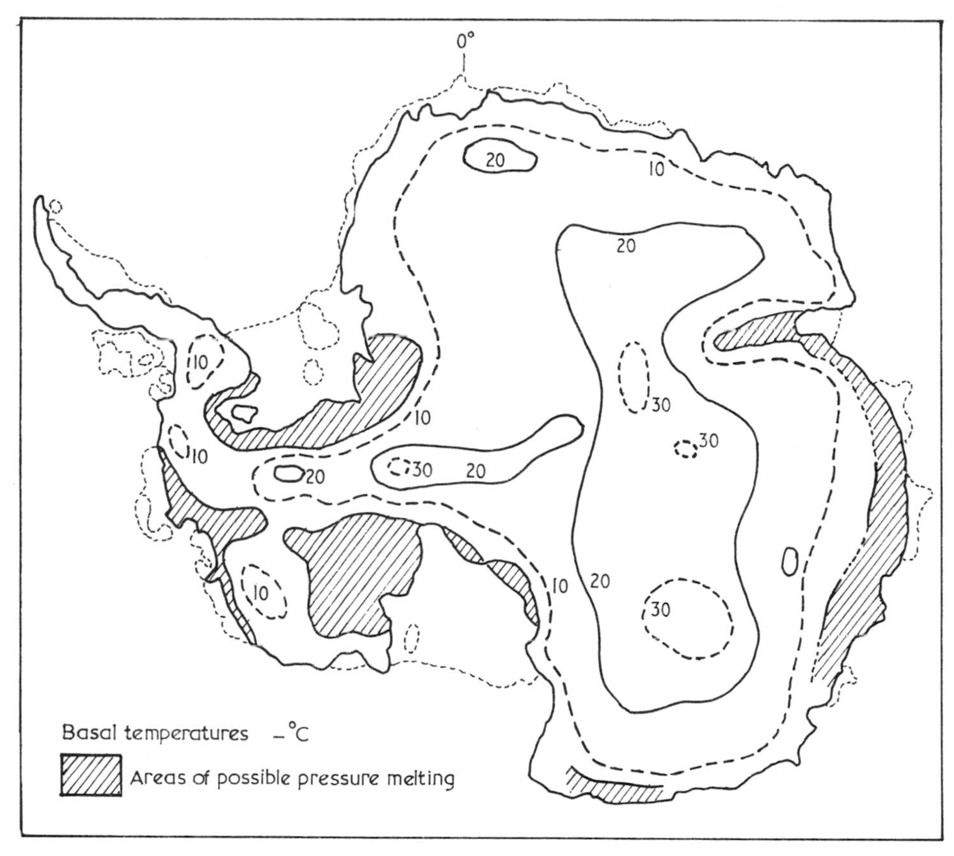

Map of the temperatures at the bottom of the Antarctic icesheet. Where pressure melting occurs, the ice can slide on the rock bed.

glaciers will be frozen to their beds so that they cannot move by sliding, movement being limited to internal deformation.

Melting of ice is influenced by pressure as well as temperature, however, and, since pressure at the base increases with ice thickness, icesheets in polar climates with extremely low average temperatures (such as the great Antarctic ice plateau) may slide on their beds if the icesheet is sufficiently thick. During a glaciation, variations in average air temperatures will thus affect the proportion of glacial movement due to sliding, by influencing both ice temperatures and ice thickness.

Movement of Glaciers

The distinction between cold-based and warm-based glaciers and icesheets is important in that distinctly different sets of glacial landforms and sediments are associated with the two types. Past glaciological environments may therefore be determined by careful reading of formerly glaciated landscapes.

The proportion of a glacier's movement represented by sliding on the bed has

(*Opposite*) One of the large outlet glaciers from the Greenland icesheet terminating in the sea. The crevassed glacier is at the top of the photograph. Note the concave 'calving bay' and the huge quantities of icebergs and smaller fragments of ice calved from the glacier.

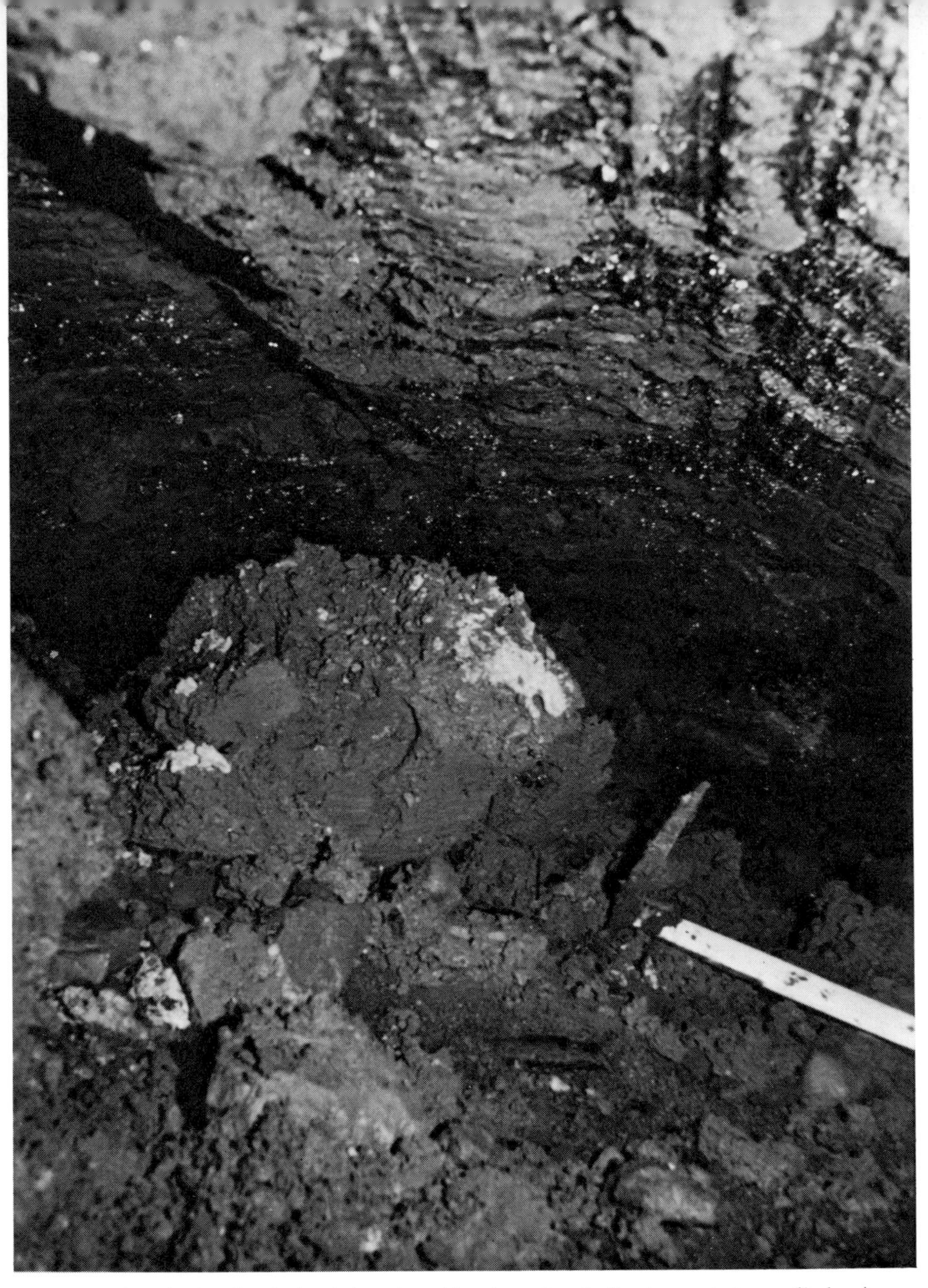

The base of an Icelandic glacier (*above*) and its deposits of silt, clay and stones *(below)*. Here a subglacial cavity has made observation possible, but most glacier beds are impossible to study directly.

important implications for the type of landscape produced. As glaciers slide they abrade rock surfaces. This may seem unlikely, as ice is 'weaker' than rock and, in any case, theory suggests that significant pressure where ice rests on a rock surface will produce a film of meltwater between the two (this is the same principle as allows skating: the high pressure between the ice and the thin blade of the skate momentarily melts the ice to create a thin film of water, on which the skater moves). The bottom surface of a glacier (the sole) is difficult to study for obvious reasons, but the limited amount of direct observation that has been

Faceted and striated stone from the Permian tillites of Victoria, Australia. The stone is just under 20cm long.

undertaken has revealed that the sole of many temperate glaciers is coated with fine debris, mainly clay and silt. This debris, of thickness rarely more than 10 or 20cm, is interspersed with pebbles, larger stones and even boulders. The concentration of debris at the sliding surface provides the conditions necessary for the abrasion of both debris and rock. In other words, rock fragments are the grinding tools at the glacier sole, which acts like a giant sheet of abrasive paper. Constant interaction between rock particles results in the production of flat or gently curved surfaces on glacial stones, which are typically scratched (or striated). Pavements of striated rock are very common in both modern and ancient glaciated landscapes, the average direction of the scratches indicating the dominant direction of ice movement.

Very high concentrations of stress may occur for short periods on the rock bed beneath a rock particle held in the glacier sole and a series of curving concentric friction cracks may develop. Where boulders in the glacier sole maintain abrasion over a longer distance, the rock may become scored into a series of glacial grooves parallel to the ice movement which created them. While the length of most grooves is measured in metres, several are known which are kilometres long and tens of metres wide. Their surfaces are often striated and friction cracks may be present.

Removal of large blocks of rock is also known to occur beneath glaciers and this raises a problem, because the strength of even quite soft rocks is many times greater than the strength of glacier ice. The answer seems to be that a glacier

Ice-smoothed rock surface near Søndre Stromfjord, West Greenland. In addition to smoothing and striating the rock, glacier ice has also 'plucked' out several large blocks of rock, as indicated by the areas in deep shadow.

may create a high stress on one part of a rock surface but, at the 'downstream' end, stress may be low or absent. The stress imposed on the rock by the glacier can be relieved only in the downstream direction and, over time, fractures develop in the rock, making possible the removal of blocks even by the relatively weak bond between them and the glacier sole to which they have become frozen. A distinctive form produced by this process (*plucking*) is the '*roche moutonnée*'—so-called because an eighteenth-century observer fancied that a swarm of them looked like a group of wigs slicked down with mutton fat!

At least some periodic sliding of the glacier on its bed appears necessary for the production of glaciated pavements and *roches moutonnées*, as for many others of the landforms considered typical of glacial erosion, some of which we will briefly mention.

As we have seen, the growth of glaciers in the middle latitudes begins in the high mountains, and it is here that glaciers remain longest as the climate improves with the approach of an interglacial period. The high valleys in many mountain regions have been given a distinctive scale and shape as a result of prolonged glacial erosion.

At their upper ends are cirques, the characteristic erosional landforms of glaciated mountains. The exposed upper surfaces of a cirque (the *headwall*) are sharp and angular as a result of frost shattering: they clearly contrast with the smoothly rounded and symmetrical forms of the lower slopes which were beneath the ice.

A magnificent roche moutonnée in the Stockholm Archipelago, Sweden. The ice travelled from right to left, shearing off several slabs of rock on the lee side of the rocky hillock.

Small cirque and cirque lake in the uplands of the Falkland Islands. Note the steep headwall and the smooth 'lip' which has been overridden by ice leaving the cirque basin.

At the same time, as the ice moves along its valley it carves a distinctive channel, the typical sharp or V-shaped form of the river valley being replaced by the broad U-shaped cross-section typical of mountain glaciation.

The ability of glaciers to create hollows by erosion of solid rock ('overdeepening') results in lakes of rockbasin type. Such lakes, together with those dammed up by glacial debris, are typical of formerly glaciated regions: it has been estimated that 90% of the world's natural lakes are the product of glaciation during the Pleistocene.

When ice masses grow to icesheet proportions, the zone of erosion spreads outwards from the mountains onto the lower lands. Given a thick and vigorous icesheet this zone of erosion may extend over enormous distances; the two greatest icesheets of the Pleistocene, the North American and the Scandinavian, have left vast areas of landscape made up of scratched and polished rocks with hollows between them on which there is little or no soil. Similar wide expanses of glacially eroded rock are also associated with the earlier ice ages of the geological record.

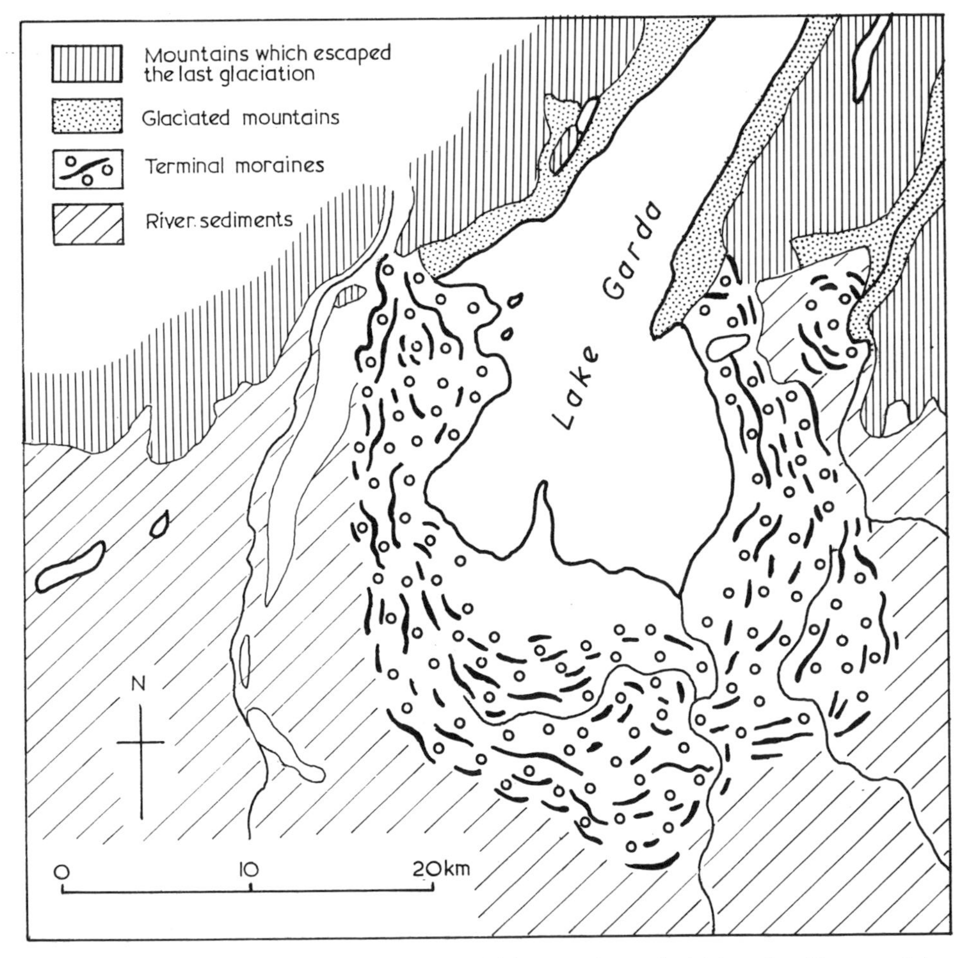

Lake Garda in northern Italy—a typical lake which occupies a glacial trough with a morainic 'dam' at its exit.

In high polar regions, in which all but the thickest glaciers and icesheets are frozen to their beds, sliding may not occur. One effect of this is to protect rock surfaces from the glacial grinding and also from the ravages of atmospheric weathering. Many of the thinner ice bodies around the periphery of Antarctica appear to be frozen to their beds and to be effecting little or no glacial erosion. In the right conditions, therefore, a landscape may become covered in glacial ice but suffer no significant modification.

While it is theoretically possible for a glacier to develop and leave no trace of its former presence in the landscape, in practice this is a remote possibility because of the changing climate and the resulting change in the state (or *régime*) of the glacier, especially during its waning phases. Many of the glaciers currently frozen to their beds in the McMurdo Sound region of Antarctica are set in a landscape made up of typical glacial forms, leaving no doubt that for prolonged periods these ice bodies moved by sliding as well as by internal deformation. However, this may have been several millions of years ago. In other words, although they are cold-based under the present climatic régime, they have been warm-based at times in the past.

The change from movement by sliding to movement by deformation may occur in space as well as over time. The transition from the thin marginal ice zone which is frozen to its bed to the thicker interior ice zone undergoing basal sliding is of major importance, whether this persists for centuries, as in polar regions, or merely through each winter, as in many mountain glaciers and ice caps. It is important because it provides a mechanism for detachment of rock debris in the process of glacial erosion. A zone of high stresses may develop in that part of a particular glacier where the actively sliding, thicker ice gives way to the non-sliding, cold-based ice in the marginal zone. Stresses may become so high that they are relieved by one or more fractures which allow the thick ice to ride up over the inert ice in front. These shear planes provide one mechanism by which rock debris from the base is lifted towards the surface of the glacier. Once they have developed, shear planes may continue to relieve internal stress because the layer of silt and clay lining them is weaker than the surrounding ice.

In suitable conditions, large rock units may be plucked by the glacier. For example, where a cold-based section of a glacier rests on permanently frozen ground, the bond at this surface may be stronger than at the transition surface between frozen and unfrozen rock. With forward movement of the glacier, stress may be relieved within the rock rather than within the glacier. It is this process which has been used to explain the presence of huge rock slabs within glacial sediments—the so-called 'giant erratics'.

Glacier Power

The factors affecting glacier mass balance (see page 62) influence not only the distribution of glaciers but also their degree of activity. In very dry polar climates and on some very high continental mountain areas, such as those in central China and northern Tibet, glaciers may develop despite relatively low snowfall amounts; rates of ablation are also very low, however. These glaciers are said to have a low 'activity index', and they achieve comparatively little erosion. In contrast, glaciers are maintained in maritime climates despite high rates of summer ablation because the winter snowfall amounts are so very

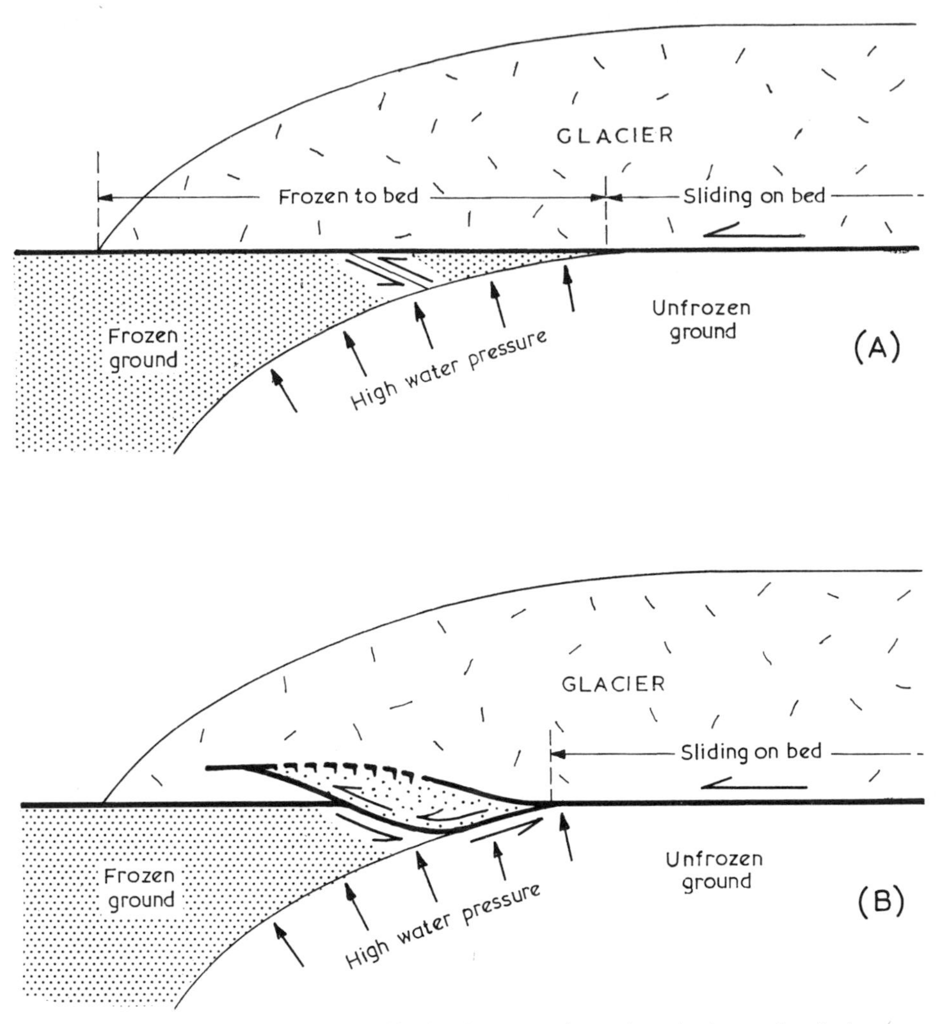

Diagram showing how large blocks of bedrock are caught up into the base of a glacier which is frozen to its bed. Failure of the rock (A) occurs where the 'sliding' part of the sole gives way to the 'frozen-on' part. The movement of the glacier, combined with the high water pressure beneath the frozen ground, causes the slab of bedrock to be pulled away (B).

heavy; glaciers of this kind have a high activity index and a high potential for erosion.

Such variations are important because the work a glacier can do in a given period of time—that is, its *glacier power* (expressed in its erosion of the landscape and in its transporting and subsequent deposition of debris)—is greatly influenced by the stress it applies on its bed and the velocity at which it is sliding. This power is greatest in glaciers most of whose movement is by sliding on the bed (up to 80% of the movement of warm-based glaciers may be by this mechanism). Polar glaciers which are frozen to their beds, on the other hand, have very low values of glacier power. In fact, sliding velocities and hence glacier power in warm-based glaciers, such as those in Iceland or Norway, may be ten or more times greater than in cold-based glaciers.

All this suggests that the rates at which glaciers erode the land surface vary

Subglacial stream emerging from beneath the Franz Josef Glacier, New Zealand.

according to whether they are warm- or cold-based. Rates of glacial erosion are, of course, not easy to measure: access to a large area of the bed of a modern glacier is not a practicable proposition! What we know about erosion rates is therefore in the form of estimates based on indirect methods.

One useful approach is to measure the debris carried in the streams of meltwater emerging from an ice front. If we estimate the total amount of the stream's load in solution, suspension and that being dragged along the stream bed over a whole glacial basin for a specified number of years, the rate of lowering of the land surface by glacier action may be estimated. Another approach is to compare the probable surface shape of a landform such as a valley before glaciation with its present form. An estimate can be made, for example, of the volume of rock removed in transforming a river-cut V-shaped valley into a typical glacial U-shaped valley; if this is divided by the time over which glaciation persisted, an estimate of erosion rates is provided. The results of calculations based on both of these methods suggests that between 0.5m and 5m per thousand years may be representative for warm-based glaciers.

Cold-based glaciers pose a problem, however. When meltwater is produced by these glaciers a good deal of it is concentrated on rock and debris surfaces along their margins. Yet other polar glaciers may have a significant amount of summer meltwater derived from streams on the surface of the ice. Comparison of reconstructed pre-glacial surfaces into which have been cut the relatively shallow cirques of more continental and polar areas suggests that rates of glacial erosion under such conditions may be as low as 0.05m in 1000 years.

The figures given here cannot be more than very crude estimates, however, especially when it is recalled that both cold- and warm-based conditions may prevail in a glacier many times over during the passage of a glacial period.

Glacial Deposition

Glaciers pick up rock debris from a variety of different sources. Most comes from the glacier bed and the glacier sides where moving ice is in contact with the bedrock, but rock fragments can also fall onto glacier surfaces from steep valley sides or from frost-shattered mountain peaks. The debris may be carried beneath the glacier (subglacially), within the glacier (englacially), or on the surface (supraglacially), before it is eventually deposited. Beneath the glacier the detachment of debris from the bed is possible, as we have seen, even though part of a glacier is frozen to its bed. Given time, the shifting of the critical point on the bed where sliding gives way to freezing-on will vary because of changes in mass balance and hence in the glacier flow. Thin polar glaciers which are frozen to their beds throughout contain very little debris. As a result their deposits tend to be thin and scattered. Temperate glaciers, as we have seen, may transport great volumes of debris on their surfaces. However, as they are wet-based throughout, they tend to have very little debris within them and, even in the basal zone, debris is highly concentrated but its total volume is small. Between these two extremes are the sub-polar glaciers. They tend to be below the pressure melting point in their upper layers but above it elsewhere so that the inner parts of the glacier move by basal sliding and a broad outer zone is frozen to the bed. Subpolar glaciers in Spitsbergen contain great quantities of debris in these outer zones which become highly concentrated as the ice melts out. Some research workers believe, therefore, that glaciers of subpolar type of produce the greatest volumes of ice-eroded debris.

The movement of large glaciers, especially those in which both warm-based and cold-based zones exist, results in the disturbance of the rock and soil beneath. With movement, and grinding of one rock particle on another, the material breaks down into various size ranges, so that with a mixture of rock types involved, the resultant deposit is a very mixed one indeed. Typically it consists of boulders and pebbles set in a fine matrix of sand, silt and clay. It is mixed not only in the sizes of the fragments but also in the minerals and rock types it contains. This is *till*, the typical product of glacial deposition.

Tills may be laid down beneath the glacier or may melt out on its surface ultimately to flow off slowly to form a rather irregular sheet. The zone of concentrated debris at the base of many glaciers may be released as a result of

Thin polar glacier containing very little debris: Taylor Valley, Antarctica

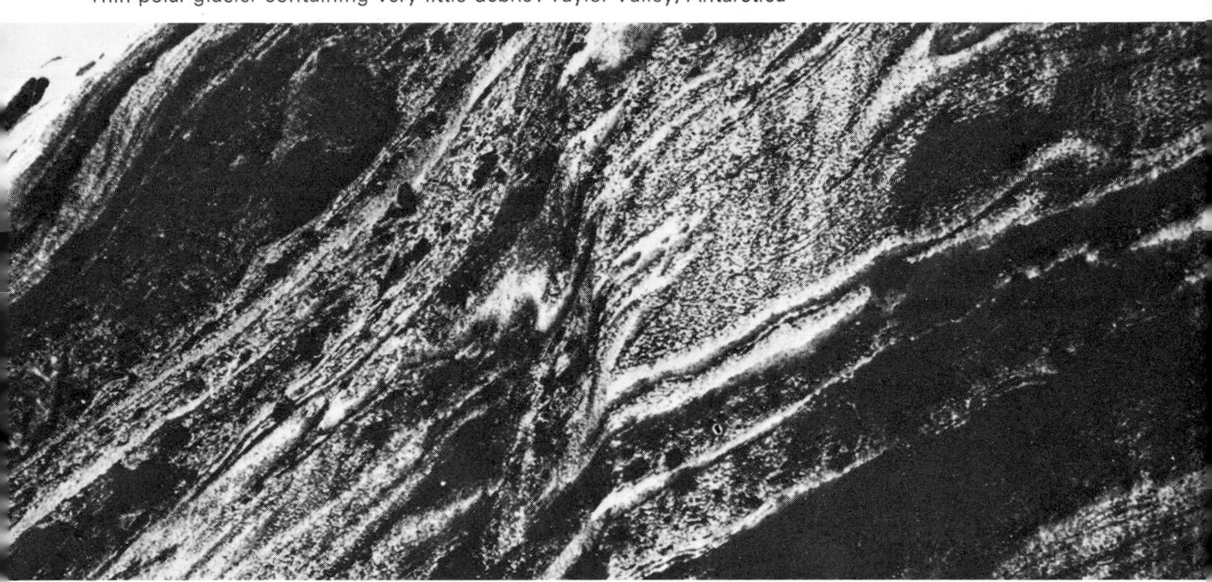

Part of the snout of a subpolar glacier in the South Shetland Islands, Antarctica. The ice layers are highly contorted in places, and there is a great deal of rock debris in the ice.

Part of the surface of a glacier in Spitzbergen. The 'face' of melting ice is covered with rock debris, much of it derived from the surface layer of ablation moraine as the ice melts back.

Broken-up rock surface affected by a small glacier in northwestern Iceland. Rock fragments such as these are the raw materials from which till is made.

Stony till of Pleistocene age in the Arm River Valley, Tasmania.

melting of the enclosing ice by frictional heat generated by the glacier's sliding, by the flow of earth heat from below and by pressure melting. Such deposits, laid down under relatively high confining pressures, are commonly rather massive in appearance and very dense.

Concentration of boulders at the surface of a till by the sluicing away of the finer material may be followed by a period of erosion, in which the boulders are truncated and polished to produce a boulder pavement. After this, further deposition may take place so that the pavement is buried, thus preserving, within the till, a record of the ice movements. The presence of boulder pavements within tills of Quaternary age is rare, but they have been reported from the Dwyka Tillite in South Africa.

A great deal of till is deposited under normal (atmospheric) pressures. Debris falls off the glacier sole where thin ice rides over rock cavities, and this debris slowly loses its ice content to make meltout till. The same sort of process is widespread on the surface near the edge of the glacier, the debris being slowly let down so that it has a rather loose structure in comparison with till laid down beneath the glacier. In some very cold and dry polar environments, the ice content may be removed by sublimation—the direct change from ice to water vapour with no liquid intermediary stage.

Release of meltout tills on sloping ice surfaces may give rise to the sliding and flow of debris. Constant supply of meltwater may result in prolonged movement of debris in a slurry which, when finally deposited, is called flow till. Tills of this type vary in their density and frequently show a crude layering structure as a result of accumulation as a series of thin flows. The process is of considerable importance especially around the margins of modern subpolar glaciers, where thick and extensive deposits of interlayered flow till and meltout till are to be found.

Long-standing problems of interpretation of complexly interbedded tills of the Pleistocene glaciations have been resolved as a result of the realization that not all tills are of subglacial origin, and that a formation of several tills interbedded with meltwater stream deposits may be the product of a single advance and retreat of the glacier. Moreover, current work on the tills of mid-latitude Europe suggests that, although the outer areas of the great Pleistocene icesheets were polar or subpolar in type, beneath the centres of the icesheets conditions were warm-based throughout.

Tills which become consolidated and cemented until they attain strengths equivalent to soft rocks such as sandstones are called tillites. The discovery of tillites dating from Precambrian, Ordovician, Silurian, Permian and Carboniferous times in different parts of the globe is important evidence indicating repeated glacial ages, shifting of the poles, and movement of the continents (we will come back to this in Chapters 4-6). Not all tillites are very old, however. Beneath the till of the last glaciation of Tasmania, for example, is a tillite which (by the plant species indicated by pollen grains in its associated lake sediments) can be firmly dated to the later part of the Pleistocene.

The wide range of rock-particle sizes in till, from boulders down to the finest clays, sets it apart from most other sediments which are to some degree better sorted in terms of the sizes of particles they contain. However, not all bouldery clays are of glacial origin, and the following chapters refer to some heated

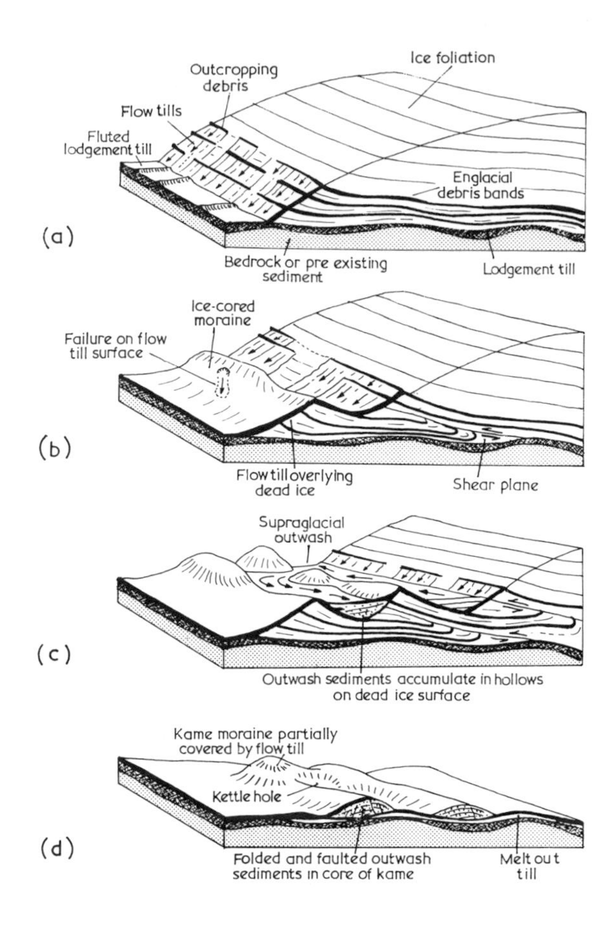

How different types of till are formed on a wasting glacier margin. Interbedded tills and meltwater deposits can be formed without any glacier readvance.

debates concerning the status of so-called 'tillites' of Precambrian and later age. We will mention a few examples here.

Ancient mudflow deposits possess all of the properties of tills to some degree, even to the extent of containing occasional striated stones; flow structures and locally well sorted layers are common, as in some flow tills.

Much rock debris of mixed grain sizes may be produced by mechanical disintegration in non-glacial frost climates. This material is moved slowly downslope over wide areas by the periodic heaving produced by growth of ice in the airspaces in the rocky mass. The result is a relatively loose structured soil called solifluction earth, which has rather angular stones reminiscent of some superglacial meltout tills.

A variety of slide and flow debris likely to be confused with tills and solifluction earths is produced by rock and soil failure in areas subject to earthquake shock. This has led in the past to claims that Pleistocene glaciers descended to altitudes as low as 350 metres above sea-level in subtropical areas such as South Central China.

Turbidites are an important and widespread group of sediments with many properties similar to those of till. Turbidity currents are suspensions which, because they are so full of sediment particles, are far denser than the surrounding lake or sea water. As a result, they move rapidly along the lake or sea floor. Some turbidity currents are sufficiently powerful to transport quite large rock fragments as well as finer-grade material. The tendency for the coarser particles to settle out before the finer ones produces layers of different particle size in the upper part of a turbidite; some turbidites, however, may appear massive and so more closely resemble tills, particularly where the upper beds have been removed by erosion.

Other sediments which might be confused with tills and tillites are derived from floating ice, which releases material from within it as it melts. In some areas a continuous 'rain' of debris of all sizes may be dropped onto the sea or lake floor: the resulting sediment may be a rather massive stony clay with little or no evidence of having been deposited in water. These stony clays may closely resemble land tills of massive type deposited on land, and have been called waterlain tills. Occasionally thin layers and lenses of sand and occasional sorting of the particles into aeras of different sizes can indicate a subaquatic origin.

Deposits laid down in this way often contain distinctive marine fossils. Sediments laid down on the floor of the Ross Sea, Antarctica, by release from a large ice shelf exceed 360m in thickness. They contain the tiny marine organisms called Foraminifera and diatoms which provide information on the environment at the time of the sediments' formation. The sediments are known to have accumulated at an average rate of about 40mm per thousand years.

When ice-shelf margins break up, the huge icebergs so released may be transported hundreds of kilometres by ocean currents. With movement into warm water, sediment is slowly released as the iceberg gradually melts away. Icebergs derived from glaciers rich in debris (such as subpolar glaciers or polar glaciers thick enough to develop some warm-based areas) provide a way in which rock particles of glacial origin can be transported over considerable distances. The occasional facetted and striated boulders found in marine rocks may owe their origin to this.

Moraines: Landforms of Debris

During deposition, tills and the sediments associated with them can be moulded by the glacier into distinctive landforms known as moraines. There are a number of different types of moraines associated with different processes of formation. We shall briefly identify and describe a few of these types although, of course, our list will be far from exhaustive.

The debris covering the surface of a glacier gives the ice protection from the heat of the Sun, and this often results in a ridge of dirty ice known as a medial moraine.

These ridges become broader towards the front end of the glacier and, at the ice front, frequently give rise to promontaries of debris with cores of ice. As these promontaries waste down, sliding and flow of the debris constantly reveals new surfaces of bare ice; the ice melts and the process continues. By the time all the ice has melted away the mass of till left behind may be very irregular in form. Such masses are called hummocky moraine.

Some hummocky moraine of Pleistocene age has a form, structure and

(*Left*) Ice-cored disintegration features (including hummocky moraine) on the wasting surface of the Hooker Glacier, New Zealand.

(*Right*) Large lateral moraine left behind by the down-wasting of a debris-covered glacier: Mueller Glacier, New Zealand.

tendency for the contained pebbles all to be oriented in a single direction, suggesting an origin due to the squeezing water-soaked till into cavities beneath the glacier or into crevasses in immobile ice. The concentration of till and scree between the glacier edges and the steep walls of its valley may be great enough for ridges several hundred metres high to accumulate. In these moraines, described as lateral moraines, the component rock fragments are often roughly parallel in orientation, and this would indicate formation of the moraine by sliding and flowing off the surface of the glacier.

Moraines running across the snout of a glacier are called end moraines (or terminal moraines). Thickening of a deposit of till beneath an active ice front, which is able to maintain its snout in the same position over a number of years, gives rise to end moraines formed by lodgement. When a glacier advances across previously deposited till, the simple 'bulldozing' action of the ice often produces so-called push moraines. Other processes which concentrate till into end-moraine ridges include accumulation of till at the ice edge by simple sliding and flowing; and the force exerted by the glacier on thick, water-soaked tills to produce small squeeze-up end moraines.

Interaction of boulders between the glacier sole and accumulating till beneath the glacier results in clustering, which often takes the form of zones of boulders

(*Opposite*) Stripes of medial moraine on the surface of a valley glacier in east Greenland. Note also the crevasse patterns on the glacier surface.

End moraines formed by lodgement at the terminus of the Hooker Glacier, New Zealand. The moraines are asymmetrical, with the gentler slope facing the glacier (*right*).

lying parallel to the direction in which the ice is moving. Such an aggregation moves much more slowly than the overriding ice and ultimately comes to a halt. The glacier flows around this obstacle so that a vault is created in the glacier sole; till may concentrate in this vault, either falling in from the ice roof or as a result of the ice pressure on nearby saturated till squeezing some of it sideways into the vault. The result, once the glacier has departed, is a series of streamlined parallel ridges known as fluted moraines. The size of fluted moraines varies from a few tens of centimetres up to several metres, according to the size of the obstacle responsible—whether it was boulders at the bed or a bedrock protuberance such as a *roche moutonnée*.

Much larger streamlined forms in glacial drift of Pleistocene age are widespread, and some are known from earlier glaciations. They are more difficult to explain because the small-scale process responsible for fluted moraines does not work if you simply have a bigger obstacle, thicker ice and higher sliding velocity; the process is ineffective if the vault is more than a few metres high because plasticity of ice near the bed increases with ice thickness, and this tends to close up basal cavities. The streamlined features of the 'half-egg' form known as drumlins appear, from the structure of the till in them, to be the product both of layer-by-layer accumulation under the glacier and of selective erosion. The latter process is clearly evident in those drumlins where 50% of their volume is made up of essentially undisturbed glacial lake and stream sediments with a surface coating of till.

Both fluted moraines and drumlins are commonly associated with ridges of till which lie at right angles to the direction in which the ice was moving. Many of these ridges look like end moraines. Their intimate association with streamlined forms, however, their remarkable regularity of spacing and their extent over great areas have led to the belief that they are formed beneath icesheets. Those known as Rogen moraines in Scandinavia are often 'drumlinized' and

Small ridges of till (fluted moraines) emerging from grooves in the base of a glacier (*foreground*). Breiðamerkurjökull, southeastern Iceland.

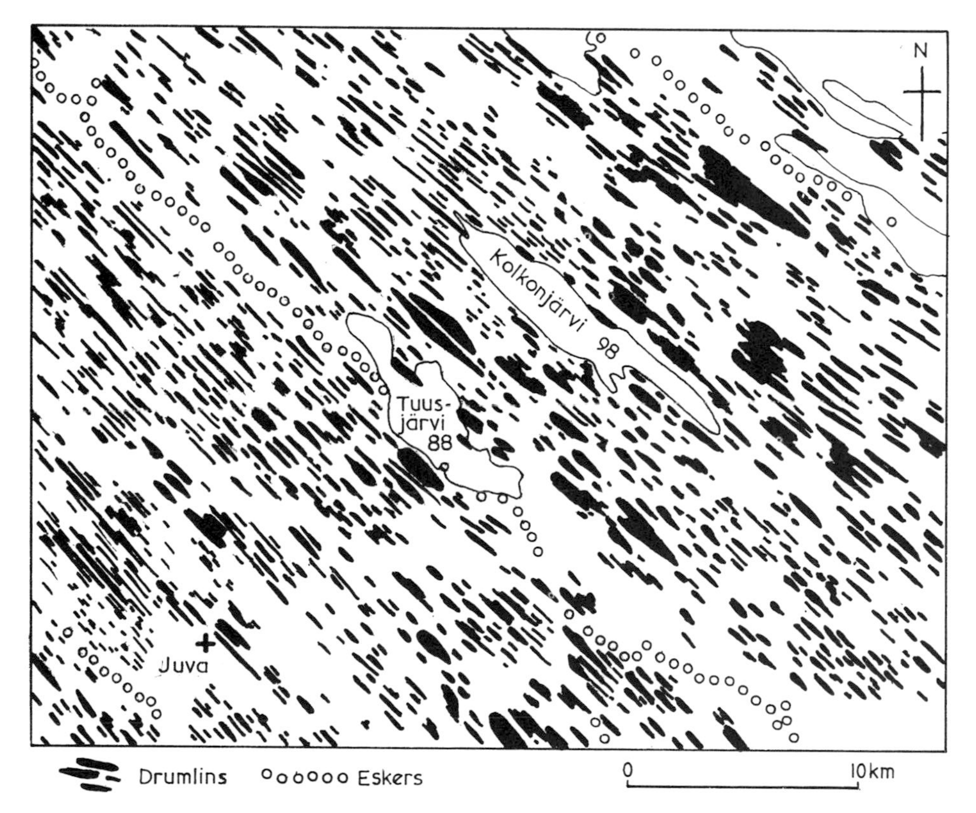

Part of a large drumlin 'swarm' in Finland. The drumlins and large-scale fluted morainic ridges clearly indicate the direction of ice movement at the time of their formation.

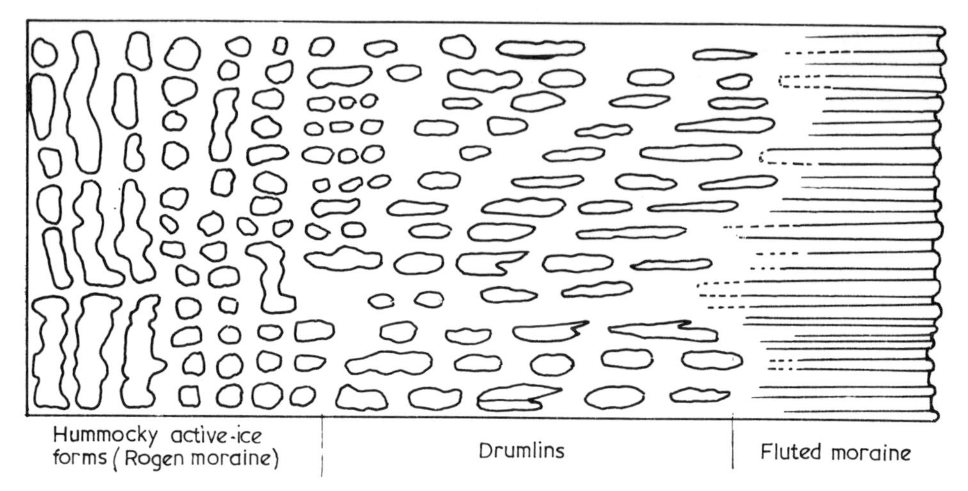

Diagram to show how various types of morainic landforms are related in part of Finland. The ice moved from left to right.

pass into a drumlin zone and then into fluted moraine. This suggests that all are types of subglacial bed forms, secondary elements of flow at right angles to the main flow perhaps being present in the basal ice and accounting for the drumlinized elements.

Another widespread type of subglacially formed moraine at right angles to the ice flow is called De Geer moraine. These are relatively small (5-15m high) ridges consisting of till, stratified sands and gravels and regularly layered lake or marine silts and clays. De Geer moraines frequently occur in broad depressions in the landscape and have long been regarded as the product of the grounding of icesheets in relatively deep water. The inner (grounding) zone of a floating glacier snout provides a zone where till is accumulated and thrust by the ice to produce a ridge. It is also a zone where there are tension cracks in the ice, the calving of icebergs from the icesheet being controlled by the spacing of these. These relationships may explain the regularity of spacing of many fields of De Geer moraines. The fact that extensive fields of moraines of this type occur in Scandinavia and Canada, for example, is important because it implies widespread formation of moraines under icesheets calving into bodies of standing water. As far as the last icesheet of the Pleistocene is concerned, both the number of ridges and the volume of sediments represented by De Geer moraines are greater than are represented by end moraines formed at the ice front. In other words, De Geer moraines arise from a major type of glacial depositional process which can only be studied using relict forms: many of the moraine-forming processes that we can study in modern glaciers may have been of relatively minor significance during Pleistocene and more ancient glaciations.

The Work of Meltwater

When the crystalline ice and the surface snowcover of a glacier melt, water is the inevitable product. Most glaciers produce some meltwater, and temperate glaciers produce impressive volumes which can be seen moving in sheets and streams on the glacier surface. Water moves also through tubes or conduits

Icebergs in a proglacial lake produced by the calving of the glacier front, Breidamerkurjökull, southeastern Iceland. Crevasses determine the sizes of the blocks which float free. Possibly De Geer moraines are formed beneath the water surface in conditions such as these.

A magnificent meltwater tunnel cut in the Werenskjold Glacier, Spitzbergen. The icy walls of the tunnel have been smoothed and polished by the meltwater which once filled it.

within glaciers and at the glacier sole as thin films and in ice-cut and rock-cut channels. The volume of meltwater which enters the glacier's drainage system is very variable, depending on the type of glacier (cold- or warm-based), the season of the year, the presence of water-bearing rocks (aquifers) in the bed, and the amount of snow- and rainfall. Moreover deep depressions in the bed beneath very thick polar glaciers can concentrate water in subglacial lakes.

The front parts of subpolar glaciers, being frozen to the bed, can hinder the emergence of subglacial waters so that the water pressure builds up until there

A large meltwater river emerging from beneath the Kaldalon Glacier, northwestern Iceland. The water is very turbulent and is heavily charged with debris.

is an eruption in the form of a violent water spout. Water is sometimes sealed up in conduits and large subglacial cavities as well as occurring in large ice-dammed lakes. If there is very rapid melting—because of subglacial volcanic action, very rapid glacier advance (surging) or if the ice tongue damming up a lake begins to float—this water can be suddenly released. Catastrophic bursts of glacial meltwater (called *jökulhlaups* in Iceland) may involve enormous volumes of water—25-100,000 cubic metres per second, say—and can cause a great amount of very rapid erosion and transport large quantities of material.

Regular floods of large volumes of subglacial water and overflow from ice-dammed lakes encourage the development of rock-cut meltwater channels. The presence of such channels beneath modern glaciers is difficult to demonstrate, but they are found in abundance in most areas covered by icesheets during the Pleistocene. The direction of slope of the original hillside—if any—on which the glacier lay is a relatively unimportant factor in fixing the direction of flow of this meltwater: subglacial channels are aligned in the direction of the surface slope of the ice from which they are derived—that is, from areas of higher pressure (thicker ice) to those of lower pressure (thinner ice). The result is a highly consistent regional pattern which reflects the general pattern of ice wastage. Sometimes channels are deeply cut into rock by meltwater flowing *uphill* in conduits which may be over 100m in diameter. Channels cut into a rock slope at a place where it adjoined an impermeable ice edge may show consistent gradients and spacing. After the departure of the glacier, most meltwater channels appear as anomalous features of the landscape—i.e., features that cannot be explained in terms of the normal forces that have shaped the local landscape. They tend to be steep-sided, their broad floors being dry or, sometimes, occupied by tiny 'underfit' streams.

In some areas the great bulk of the deposits resulting from glaciation is, in fact, the result of transport and dumping by glacial meltwaters, particularly during glacier retreat. The rate at which rock particles settle in water varies

Violent water-spout near the front of an Icelandic glacier where subglacial water forces its way to the surface under high pressure.

The impact of glacial meltwater. A glacial meltwater channel cut in the coast of Wales. Channels such as this are frequently cut by meltwater flowing uphill under great pressure.

Parallel glacial channels formed at the edge of a downwasting Pleistocene icesheet in central Labrador-Ungava. The coniferous trees show the scale.

Faulted beds of sand and gravel in a kame terrace formed at the end of the last glaciation of western Wales.

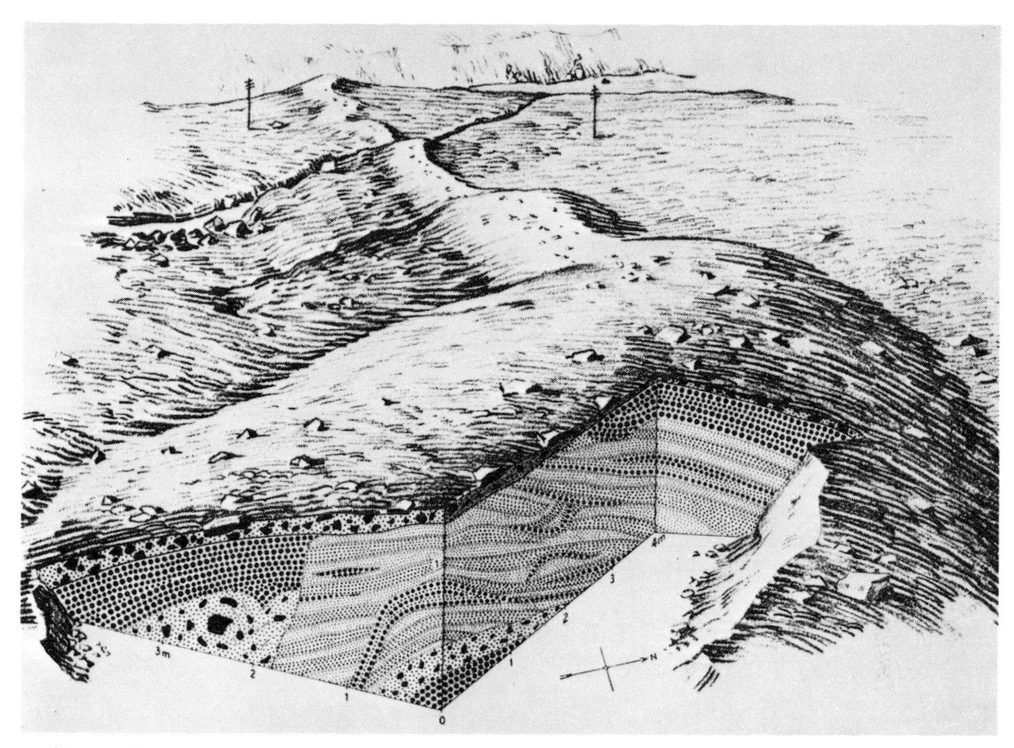

Sketch of a winding esker ridge in central Sweden, showing the complicated deposits of sand, gravel and till in the esker core.

with the size and density of the particles, with the temperature of the water, and with the velocity of the current. As a result, waterlain sediments are bedded or stratified.

The landforms produced by the accumulation of bodies of sands and gravels at the edge of a glacier, in contact with the ice, are distinctive. Bodies laid down in the angle between the valley side and the ice edge are known as kame terraces. Where outwash debris accumulates over stagnant ice, the uneven rate of ice melting causes sliding and falling of debris into cavities, the result being a series of gravel cones, short ridges and small plateaux known as kames. Streams flowing in channels on a glacier surface or in tubes within or at the bed of a glacier carry large volumes of sand and gravel. When these are finally laid down on the land surface as the glacier wastes away, a distinctively sinuous or straight ridge known as an esker may be preserved. Eskers lie approximately parallel to the slope of the ice surface, and so groups of eskers may be useful indicators of the direction in which the ice was moving.

The finer particles—silts and clays—may remain suspended in meltwater streams even when there is very little water in them, but cobbles and boulders are usually transported only when the streams are rather more torrential. The facetted and somewhat angular glacial stones rapidly become rounded at the edges as they are tumbled along the stream bed and jostled together. The result of these two sets of processes is an outwash plain, or sandur, made up of coarser material underlying finer material, across which runs a constantly shifting series of shallow but wide stream channels. Removal of fine-grained

Sandur deposits of a valley train below the Tasman Glacier, New Zealand.

material by infiltrating water or by the wind leaves a surface dominated by boulders. Fossil sandar have been recognized in parts of the world which now have a warm climate. The fossil sandar of the Sahara make up part of the evidence indicating glaciation there during the Ordovician period.

Because the shapes of glacial stones and the striations on them can be modified by the tumbling and collision in the meltwater stream, however, it is by no means always possible to be certain that sandar-like rock sequences are the product of glacial meltwater. A deficiency of silts and clays and a high frequency of cobbles and boulders is not enough, as similar sediments are to be found mantling the forelands of recently active mountain fronts in regions such as the southwestern United States and the plains of Old Castille in Spain.

When streams of meltwater enter a lake or the sea, their velocity is checked

A gigantic 'dropstone' revealed at low tide on mudflats in Scoresby Sund, East Greenland. This boulder was dropped by a melting iceberg, and is slowly being covered as marine sediment builds up around it.

and sediments are deposited. The larger and heavier particles are laid down first and progressively finer particles are deposited further out. The sands and gravels may become concentrated to produce a delta, but the silts and sands will remain in suspension in the water.

The glacier's seasonal cycle of freeze-up and melt may result in the release of pulses of glacially derived sediment into lake or ocean in summer. The coarser, sandier component is laid down relatively rapidly but the silts and aggregates of clay take much longer, and only a very thin layer may accumulate before the next summer's pulse of sands. These coarse and fine couplets are called rhythmites and, when they are strictly in annual couplets, they are known as varves. Glacial rhythmites are associated with bodies of brackish and fresh water into which a glacier debouches. When the glacier withdraws from the water body, the strictly seasonal control is removed and sedimentation in couplets gives way to more massive deposition. Thus counting glacial varves has provided a calendar of the retreat of the last icesheets in Scandinavia and eastern North America, a dating technique which preceded the advent of radiocarbon dating.

The use of rhythmites in this way is far from straightforward, however. The pulses of sediment into the water may occur in any season (an Indian summer, say). Another problem for students of ice ages is the presence of rhythmites in rock successions of non-glacial origin. In other words, rhythmites are by no means always varves and neither are they invariably glacial. What are the clues

The work of frost. Disintegration of an erratic boulder as a result of frost shattering near Breiðamerkurjökull, southeastern Iceland. The boulder was deposited by the ice little more than a century ago.

indicating a glacial origin? Rock composition should show a diversity similar to that of associated till or tillite; in addition average particle size is coarser, sand grains are frequently sharper-edged and couplet thicknesses are usually greater than in non-glacial rhythmites.

A common clue to the glacial nature of rhythmites is the presence of dropstones. Rock fragments of boulders and pebble size fall rapidly through lake or sea water when released from ice shelves or icebergs. They may become embedded vertically in the soft silt and clay sediments, or they may topple. These dropstones disrupt the layering below them but slowly become buried as the rain of fine debris continues to accumulate around them. Unlike glacial rhythmites, dropstones are not necessarily laid down in close juxtaposition with a glacier margin.

Ice in the Ground

The effects of an ice age are not limited to those areas directly affected by glaciers. Outside the landward limits of the great icesheets, themselves limited in their growth because of low temperatures freezing the margins to their beds combined with low snowfall, conditions may be severe in the extreme. Especially on the more continental margins of an icesheet, climates tend to be dry and windy with only local accumulation of snowbanks, leaving the ground surface susceptible

A large ice lens exposed on the coastline of McKinley Bay, Northern Canada. The 'cap' of sand and silt has been lifted by the growing ice lens.

to the action of frost disturbance and wind, which removes fine-grained debris.

At the height of the glacial episodes, such near-glacial—or periglacial—regions suffered a variety of disturbances due to deep freezing in areas which we would today call polar deserts. Shattering of rock surfaces by freezing, cleaving of the ground by the growth of ice wedges, and the removal of the finely shattered material by wind are typical characteristics. Even in regions which today are maritime and temperate (such as western Europe and the eastern United States) evidence of the development of these characteristics has been found, dating from the last glaciation of the Pleistocene.

In the permafrost zone, ice in the ground is of great importance for the creation of both landscapes and sediments. The ice is quite different in its crystal structure from glacier ice.

In high latitudes, where the ground is permanently frozen to depths of perhaps several hundred metres, ice masses or lenses create a number of unique landforms. Sheets of ice can disturb the ground surface over large areas, and, if the ice melts out, a special type of pitted or cratered terrain called thermokarst is created.

Smaller ice lenses can lift up the ground surface layers as they grow; typical landforms are the 'palsas' of Lapland and the 'pingos' of Canada and Greenland. The latter are extremely impressive features, and in the MacKenzie Delta region some of them are over 50m high and over 600m in diameter. When the cores of ice begin to melt out, pingos look like miniature volcanoes, often with small 'crater lakes' near their summits.

As mentioned in Chapter 1, permafrost areas have spectacular expanses of patterned ground, with large-scale polygonal markings extending over hundreds of square kilometres of the barren tundra. The markings result from the growth of ice wedges, which may be 3m or 4m across at the ground surface and taper to a point over 10m below. These wedges are not often exposed, but they can sometimes be seen in the banks of large rivers: on the banks of some Siberian rivers, ice wedges are spaced at regular intervals, projecting like huge white marble pillars beyond the darker silts and muds which make up the bulk of the river-bank material. As they grow, ice wedges usually disturb the sediments on either side of them, exerting sufficient pressure to bend sediment layers and also to lift the ground surface. When the wedges melt, their 'casts' are filled with later sediments, often preserving the outline of the wedge and the disturbance of the flanking layers. 'Fossil ice wedges' therefore become extremely good indicators of permafrost conditions; they have been found in deposits dating from a number of different Quaternary glaciations and indeed from the older ice ages as well.

Among the other signs of permafrost are strongly disturbed sediments with tight little folds and other features called involutions; small-scale hummocky terrain; stone stripes, circles and polygons which can form without the existence of ice wedges; and slope deposits disturbed by sliding on the wet but icy 'permafrost table' during the summer months.

Stone circles formed close to an ice-cap margin in northwestern Iceland. Although there is no permafrost here, patterned ground is widespread.

In the dry areas around many icesheet margins, the finer-grained particles may be selectively removed from barren glacial formations by periodic strong winds. Sand grains are moved a few centimetres only but they may groove softer types of rock, facet and polish the upper surface of rocks, and lead to the accumulation of sheets, dunes and whaleback ridges of sand. The material finer than sand, notably silt but also including any clay particles aggregated into silt-sized lumps, is taken aloft and may be transported in suspension by the wind for hundreds and sometimes thousands of kilometres. It may accumulate to considerable thicknesses in a homogeneous sediment dominated by the minerals quartz, feldspar and mica. This distinctive sediment is known as loess. In Europe, loess of Pleistocene age is found in a great belt parallel to the former Scandinavian icesheet margins, from which much of the silt was derived; a similar situation exists in North America. Although it is to be expected that accumulation of loess on a regional scale is favoured by dry conditions and especially dry, cold winds, which are most effective at picking up and transporting particles, it is difficult to agree with those who suggest that the particles making up loess are the product of glacial grinding. Thus, while most of the very thick Chinese loess probably accumulated in the cold, dry conditions of the Pleistocene glaciations, much of it was derived from the Gobi desert as indicated by mineralogy and by the decline in the average size of the particles towards the south east. Loess older than Pleistocene has not been identified with certainty and, should ancient loess be recognized, the problem of correlation with specific glaciations would be very much more formidable than it is in the case of the Pleistocene.

Not all periglacial environments are of polar desert type. Maritime climates in which frost action is of major importance have higher humidities and greater annual snow- and rainfalls. Vegetation, albeit predominantly at ground level, may be widespread and chemical weathering of rock and debris is significant. Such an environment prevails today in northern Labrador, Alaska and parts of Iceland, for example. The dominant land-shaping process is the slow movement downhill of frost-shattered debris in the form of thin sheets and lobes of wet stony clays. The soils so produced in the Pleistocene are, while rarely of great thickness, geographically very widespread, and this must have been the case also during the earlier ice ages.

Ice Afloat

We are by now familiar with the idea that ice ages have a profound effect upon high-latitude and mid-latitude land areas. We must not forget, however, that floating ice is produced in huge quantities during a glaciation. This ice blankets large areas of the world ocean with mobile floes; it transports and deposits glacial materials sometimes thousands of kilometres away from the nearest glacier ice edge; and it affects global climate, as discussed in the previous chapter.

Many glaciers and icesheets terminate in lakes or in the sea, as shown in the photographs on pages 85 and 96. Where the water is approximately as deep as the thickness of the ice, the glacier snout will float because its density is less than that of water as a result of air and salts contained in it. Extensive ice tongues and other floating glacier snouts are known from many of today's glaciated regions,

A large tabular iceberg in Melville Bay, Greenland, dwarfing the icebreaker *Westwind*: it is over 1km long, and its bounding ice cliff is about 25m high.

notably around the margins of the great icesheets of Greenland and Antarctica. Sometimes they occur on the open coast (as in Antarctica), but more commonly they are found in deep fjords such as those of Svalbard and west Greenland. In such situations the glacier front is marked by a vertical ice cliff. The floating ice edge is affected by tides and currents, and small ice fragments frequently break off and fall into the water. In addition, larger slabs of glacier ice are detached from the floating snout, giving rise to icebergs which are carried away and dispersed far and wide by winds and currents.

Ice shelves are floating icesheets. They occur in large coastal bays in the high latitudes, and derive their ice for the most part from direct snowfall. On their landward margins they are fed by large outlet glaciers, as in the case of the Ross Ice Shelf which forms a floating extension of the Antarctic icesheet. The Ross Ice Shelf extends over an area of 803,000 square kilometres; other large Antarctic ice shelves are the Ronne, Larsen and Filchner shelves.

All shelves have very slight surface gradients, falling from an altitude of 80-100m at their landward margins to an altitude of about 30m at their seaward edge. Here the ice can be up to 200m thick, and the floating shelf edge is usually marked by a sheer cliff of ice.

Like floating glacier snouts, shelf edges are affected by tides and currents which cause stresses in the ice. Cracks appear, and at intervals vast slabs of ice shelf break off to float free in the Southern Ocean. These are the flat-topped tabular icebergs, which may be over 100km long and over 40km wide, and which sometimes survive for many decades in the high-latitude ocean environment. They are quite different from normal glacier icebergs.

Ice shelves sometimes have an important rôle to play in the deposition of glacier debris. Beneath the floating part of an ice shelf much debris may be

(*Opposite*) The floating snout of a large outlet glacier in Greenland. The glacier is breaking up into a series of large icebergs, many of which are held in the grip of ice floes formed by the winter freezing of the sea surface.

Sea ice in the Bellinghausen Sea, Antarctica. Here an icebreaking vessel is attempting to keep open a lead between two areas of old ice floes. The lead is being iced over by 'new ice'.

released by ice melting from the base or the shelf, the sediments showing some variety (including rhythmites) and a notable number of flow phenomena (turbidites). In cold-based glaciers, deposition is essentially absent until the floating tongue is reached. Extensive ice shelves may build up and, with break-up into tabular icebergs, very thick, bedded glacio-marine 'tills' may be produced (see page 91). Deposition is fairly sporadic in the outer part of the iceberg zone, so that occasional lenses and balls of till—or sometimes glacial erratics—may be produced. Nevertheless, the result of deposition directly into the water over a long period of time from ice shelves and icebergs is a rather uniform series of deposits over very large areas of the sea bed.

Pack ice is ice formed as a direct result of the freezing of the sea surface. As mentioned earlier (page 14), pack ice covers great areas in both Antarctic and Arctic regions, forming and thickening each year but also being dispersed by winds and currents. New ice in any given year is seldom more than 20cm in thickness, but over the years old ice can grow to thicknesses of 4m or more. When the pack ice breaks up into individual ice floes there may be cracks and wider areas of open water well within the pack ice belt, but when floes are brought together by winds and currents the collisions between floe edges can create 'pressure ridges' of broken ice which may stand above the pack ice surface by 10m or more. Other projections of broken ice extend downward from the pack-ice base into deeper water. Old pack ice may have a very tortured and chaotic surface as a result of the oft-repeated splitting, collision, reforming and fusing of old and new floes.

Pack ice can affect shorelines in a variety of ways, especially when floes are piled up on a beach by onshore gales. On the other hand, this type of ice does not have a great effect on sea-bed sediments since it carries little or no rock debris. Its main environmental effect is in reducing the light which penetrates the oceanic waters; certain species of Foraminifera and other organisms are therefore characteristic of the pack-ice belts—just as other species are excluded. In ancient sea-floor sediments which contain the remains of these organisms clues can therefore be obtained about the presence or absence of pack ice at the time of their formation.

Changing Sea Levels

The growth of continental-size icesheets and the expansion of mountain glaciers into ice caps greatly increase the proportion of the Earth's water locked up as ice (at the moment, over 80% of the Earth's fresh water is ice). As a result, the volume of water in the ocean basins declines and sea levels fall throughout the world. This effect can be impressive, many estimates of sea levels during the last icesheet maximum of the Pleistocene (about 18,000 years ago) being about 120m below that of the present (this is discussed further in Chapter 7). The detailed geography of the world's coasts was greatly changed by events of this kind: large areas of the continental shelf became dry land, including the southern North Sea, the Arafura Sea and the Persian Gulf. This, in turn, affected the climates of the continental margins. As we saw in the last chapter, the further from the sea any particular location is, the greater the temperature range it experiences, and the amount and distribution of rainfall is also modified. This naturally has an effect on the landscape.

A few moments of thought suggest, however, that such a simple picture of the relationship between icesheet growth and sea levels is likely to be complicated by other factors. Perhaps the most important of these are the elasticity and viscosity of the Earth's crust and mantle. Concentrations of load upon it, whether of glacier ice or of ocean or lake water, make the crust 'sag' beneath the load itself and rise slightly just beyond the edge of the load. This response of the crust to loads is known as isostasy. World glaciation thus influences sea levels not just simply by reducing the available liquid water, as we have seen; upon this pattern is superimposed another due to depression of the crust beneath icesheets and its uplift beneath oceans as the load there is reduced owing to the decrease in water present. It has been estimated that the effective sea level rise of 50-60m represented by the volume of water which melting of the present Antarctic icesheet would liberate would be decreased by one third by the depression of the crust the additional water would cause. Local differences in such response may be seen in formerly flooded marine and lake basins in the form of abandoned shorelines which are warped, subsidence having been greatest in the centre of the basin and least on the periphery.

Even quite small icesheets produce a measurable isostatic response. Since the disappearance of the last Scandinavian icesheet, the crust in the Gulf of Bothnia has risen in response to unloading by over 300m, and the response to the

Old shorelines on the margin of a lake in South Georgia. When glaciated areas are relieved of their ice load, isostatic recovery leads to the uplift of the land surface; raised shorelines are a result of this uplift.

disappearance oft he small Scottish icesheet of the last glaciation exceeds 14m.

The interplay of these three effects has important implications for the student of ice ages. The gradation of marine sedimentary types or facies (coarse pebbles and cobbles of the storm beach, sand of the intertidal zone, then silts and finally the finest clays of the depths) becomes superimposed and partially eroded with advance and retreat of the sea. With a fall of sea level, the base level of the rivers is lowered, stimulating deep cutting of their valleys and the partial removal of estuary and floodplain sediments, leaving paired terraces such as those of the lower Rhine, Thames and many European rivers. Shoreline sediments are stranded out of reach of the tide. With a rise in sea level, the lower reaches of rivers become drowned, and estuary and marine clays, silts and sands may be deposited. On low-lying coasts, fen and marsh sediments are deposited and coastal peat-growth increases. As well as laying down a complexly interbedded sequence of such deposits, changing sea levels also result in erosion, sorting and redistribution of glacial and periglacial sediments of the preceding glaciation. Borehole records from low-lying coasts on the margins of marine basins such as the North Sea provide an impressive summary of the interrelationships of the sea, rivers, glaciers, wind and frost as sedimentary agents throughout several glacials and interglacials of the Pleistocene. Study of the changing sea levels of the Pleistocene has proved richly rewarding in the light it has thrown on the mechanism of cooling and warming of each hemisphere. However, methods using marine and coastal sediments to determine the nature and development of ancient glaciations are at present crude at best.

Glacial and Interglacial Life

The waxing and waning of icesheets and glaciers in response to climatic shifts has a dramatic impact upon living organisms both on land and in the sea.

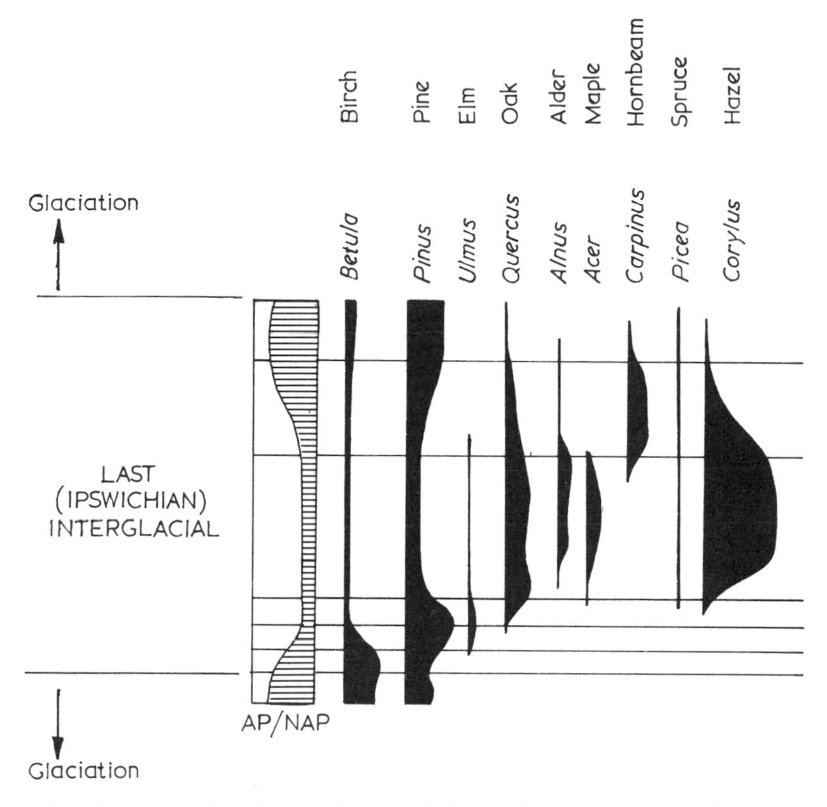

Diagram showing how pollen frequencies vary during an interglacial period—in this case the last interglacial of eastern England. The column on the left shows the proportion of arboreal or tree pollen (AP) to non-arboreal pollen (NAP). Note that tree pollen is most common during the warmest part of the interglacial.

Migration and extinction are the two obvious responses of plants and animals to severe and rapid environmental change. The third possible response, physiological adaptation (leading to the appearance of new species), is evident from the Quaternary only in certain lifeforms (such as some small mammals), almost certainly because of the relative brevity of this period.

The ability of many species to migrate during shifts from glacial to interglacial is recorded by the presence of certain types of sub-fossil evidence. The pollen grains shed by plants have proved to be a major source of information on changing vegetation patterns of the past, the relative proportion of the different species represented providing a remarkably sensitive indication of levels of warmth and moisture. Preserved twigs, leaves, bark and seeds of plants often reinforce the pollen evidence. Fossil bones of warmth-loving species such as *Hippopotamus* have been found in Pleistocene interglacial sediments in central England, while skeletal remains of both large and small mammals such as woolly mammoth and lemming are associated with glacial and periglacial sediments. Remains of several types of beetle which today inhabit northern Siberia have been found in deposits of the last glaciation in western Europe.

Oscillation between cool and warm conditions is indicated also by marine lifeforms. In the Mediterranean and the littoral of Western Europe, the decline in sea temperatures of the shallow seas is indicated by a relatively rapid increase in northern shelly species (molluscs), which has been used as a way of marking the boundary between the Pliocene and the Pleistocene.

Lifeforms of the deep oceans also reflect the changing environments which produce growth and wastage of icesheets. Of particular interest and value are the minute lifeforms known as the Foraminifera. Comparing the number of warmth-loving species present to the total number present provides a sensitive indicator of change in ocean temperature levels. Some Foraminifera species change from a growth form, during warmer conditions, in which shells are coiled to the right, to a left-coiling form in cold conditions, providing an example of adaptation to temperature rather than migration or extinction.

As stressed in Chapter 1, world glaciation, whenever it occurs, causes sweeping changes in the distribution of life both on land and in the sea. Some examples of the interrelationships involved for the Pleistocene may serve as a general case, although we can probably assume that similar relationships existed during *all* of the ice ages recorded in the geological column.

The glaciated zone during the cold episodes of the Pleistocene was greatly extended, in the northern hemisphere reaching as far south as 50°N (and even to 40°N in North America), in comparison with an average southerly extent of only 77°N today. Outside the glacial area, the subpolar or periglacial zone of tundra and steppe vegetation reached into central Italy, in its turn pushing the evergreen Mediterranean forest into a very narrow zone between 37° and 35°N.

But this simple equatorward pushing of zones could not go on if the energy and water balances of the Earth as a whole were to be maintained. The dynamics of the atmosphere are such that the zone of great heating over the equator is one of rising air masses compensated for over the deserts by descending air masses, so that the great heat engine of the atmosphere (moving heat from source areas into sink areas) remains in balance. A fundamental shift in this balance occurred, however, as the areas of polar cold expanded into what are now warm temperate or even subtropical areas. Events in tropical latitudes are not easy to interpret.

On a world scale, the locking up of so much moisture in the great icesheets, and the consequent reduction in world sea-levels, effectively enlarged the sizes of the continents and so rendered them drier. Sand dunes dating from the glacial maxima have been found below present sea-level and great fields of dunes appear to have extended on the equatorward side of the subtropical deserts; although the polar sides of these deserts were wetter, as indicated by the evidence of high water levels in the lakes indicates.

While world temperatures in all zones were somewhat below those of the present, the degree of cooling varied from place to place. The atmosphere over western Europe was probably 7 or 8C° colder than the present; about the tropics it was some 4C° below the present; but in the equatorial rainforest it was no more than 1-2C° below the present. Such minor changes in the equatorial forest zones had no perceptible effect on land processes there and no effect at all on lifeforms: indeed, in this forest zone pre-Pleistocene conditions were perpetuated such that it became the last refuge for the great Tertiary animal and plant populations. Ancient stock such as elephants, lions and giraffes were able to survive in the equatorial Old World until the present.

Further north, things were more complicated. What today are deserts and

Ice-age survivals in the tropical zone. Animals such as the elephant, giraffe, rhinoceros and hippopotamus have managed to survive the coldest episodes of the Cenozoic ice age because temperature changes are least marked in the tropics.

Mammoth skull and part of the body found on the bank of the Berezovka River in Siberia in 1902.

Ice-age extinctions in the middle and high latitudes. Animals such as the woolly mammoth, the giant deer and the sabre-tooth tiger became extinct because they could not cope with the great environmental changes of the glacial stages.

savannahs contracted slightly or shifted equatorward. Water levels in the lakes were lower than the present during the glaciations but became very much greater than they are today in the early postglacial period, as is evident in the abandoned shorelines above Lake Chad, the East African and western North American lake basins and the Caspian Sea.

Within the tropical and subtropical zones the glacial periods coincided to some extent with so-called pluvial (rainy) periods of greater rainfall and higher humidities, but to correlate them directly is a gross oversimplification. Pluvial episodes may not have lasted throughout the time of the glaciations of the northern continents, and may have been altogether briefer, higher-frequency events. The relationship to glaciations is by no means universally agreed.

As we have seen, the Mediterranean-type forests moved southwards to encroach on the northern-hemisphere deserts from the north, but the Mediterranean zone itself was probably drier during maximum glaciation than it is today. Northward of this zone (in fact northward of about 45°N) the climate, while colder, was less humid: the legacy of sediments and soil structures indicates a great extent of polar deserts.

A further complicating factor during the glaciations of the present ice age was due to the geography of the continents. Southward migration of plant species ahead of the advancing icesheets met no substantial obstacle in North America, for example, where the mountain ranges are aligned north-south and the great central plains are open and wide. In Europe, however, the situation was quite different, because the principal alpine regions are disposed east-west; as the icesheets advanced southward so the zone of migration for plant and animal species progressively narrowed. Moreover, the valley glaciers of the Pyrenees and the Alps coalesced locally to form icesheets, so that the plant and animal populations in central Europe were embraced in a kind of pincer movement of annihilation. The result has been the extinction of a great many species which thrived in Europe in the Tertiary but did not survive the Pleistocene, although they did in North America; a well known example is the magnolia.

A reconstruction of the sabre-toothed cat, one of the predators of the Quaternary ice age which became extinct before the end of the last glaciation.

Another result is the greater richness of plant species on the southern margins of the European continent (for example, in southeastern France) in comparison to the limited number of species in the forests and woodlands of northern Europe. Yet a further result has been some remarkable plant and animal distributions. Several species of heather, for example, are found in the British Isles only in the extreme west of Ireland: one has to travel all the way to the western Pyrenees to find them again.

While on the subject of plant and animal species and their distributions, we might revert to yet another result of sea-level changes. Low sea levels reduced the distance from Asia to Australia by two-thirds and greatly broadened the bridge between Asia and Africa in the Suez area several times during the Pleistocene. Asia was linked to North America through the Bering Strait and the British Isles was linked to Europe. Migration of animals was made easier, especially during the pluvials when the desert zones were narrowed in areas such as Suez: Man was one of these migrant animals. So it is that the great animals of the Asiatic mainland, such as the elephant and the tiger, are found also on the small islands of the Indonesian archipelago.

In our discussion of the effects of ice ages we have tended to concentrate on the Pleistocene ice age—quite naturally, since it, being the most recent, is the one that we know most about. The individual effects of the Earth's ice ages—including the Pleistocene—will be discussed in more detail in later chapters of this book; however, it is worth pausing briefly here to recap on some of the major effects of this particular ice age on life as we know it today.

While icesheets never invaded the tropics during the Pleistocene, the disturbance in the moisture and heat balance of the whole Earth was such that the periods of ice advance in higher latitudes were times of marked climatic and geographical change in the tropics. As parts of the tropical deserts became more habitable, with more pastures and watering places, and hence animal life, the hostile barrier which they had represented was much ameliorated. At the same time, some of the ocean barriers were reduced or eliminated altogether. It is with this kaleidoscopic background of climatic and environmental change that Man's ancestors first appear on the scene. The Pleistocene ice age was early Man's milieu, and during that ice age he became recognizably human.

The influence on plants and animals of geographical changes such as uplift and realignment of mountain chains, generation of lake basins by crustal movement, and the creation and breaching of land bridges as sea-levels fell and rose must have been of equal importance during the glaciations of the Ordovician and earlier times but, not surprisingly, analytical work has yet to reach the level achieved in the study of Quaternary changes of this sort. Nevertheless, as we shall see later in this book, broad climatic zonation has been recognized, on the basis of fossils, around the margins of the African Ordovician icesheet; and mountain ranges dating from the time of the Permo-Carboniferous glaciations have been recognized in northern Argentina and Bolivia and in central Patagonia with associated basins of deposition, containing glacial and other sediments.

Most authorities now accept that the course of evolution has been greatly altered by the present ice age. It is therefore logical to assume that the early ice ages must have made a similar impact upon the evolution of plant and animal species, the make-up and distribution of plant and animal communities, and the extinction of species which were not capable of either migrating or adapting to periods of low temperatures and environmental changes. Some of the points made in the preceding chapters are relevant here, and in the chapters which follow we demonstrate over and again the overwhelming importance of ice ages in the explanation of our planet's web of life.

EDWARD DERBYSHIRE

4 The Earliest Ice Ages: Precambrian

The Precambrian eon represents about 85% of geological time: it lasted from 4,600 to 570 million years ago. Throughout this vast timespan, unequivocal evidence of widespread glaciation on Earth is recorded only during two broad periods. The oldest ice age occurred during Lower Proterozoic times, and its deposits are best preserved in North America. A much younger and more extensive series of glacial events, which may or may not have been separate ice ages, occurred in Upper Proterozoic times, ending just before the onset of the Cambrian period. Before considering these glaciations and the record that they left on Earth, let us begin with a brief discussion of the conditions that prevailed on our planet in these early times.

Early Environments and Rocks

The Precambrian world would have been largely unrecognizable to us. In the first place, we would not have been able to survive in it, since the atmosphere contained only a minute amount of oxygen. For much of the time, at least during the early part of the Precambrian, the Sun was hidden behind unbroken banks of cloud. It rained—possibly for millions of years—without a break. Everywhere there were volcanic eruptions with outpourings of lava. As the crust and the atmosphere evolved so did the ocean. It must originally have been a real 'world ocean', covering virtually the whole of the Earth's crust. Eventually, as great thicknesses of lava and other volcanic materials accumulated in some areas, the early continents were born. Mountains, basins and plateaux may have been created by primaeval tectonic forces. Rocks were broken down by early processes of weathering which may have been rather different from those which operate today. The land surface was subjected to attack from gravity, wind, water and (in the polar and high mountain areas) ice. Gradually valleys and hill-slopes were carved out of the landscape and rock debris was redistributed, carried from the higher areas and deposited mostly in lakes, on valley floors and in the ocean.

At sea there was no life as we know it. On land there were no plants and no animals. For millions of square kilometres there stretched lifeless landscapes—undulating expanses of volcanic rocks with scattered pools and lakes, sandy deserts, landscapes of volcanic craters and grotesque lava formations. The sounds were the primaeval ones, coming from volcanic eruptions, rockfalls and avalanches, thunder and wind, incessant rain and running river water. On the seashore there were the sounds of crashing waves against cliffs and on sandy beaches.

The Precambrian environment would not have struck us as hospitable.

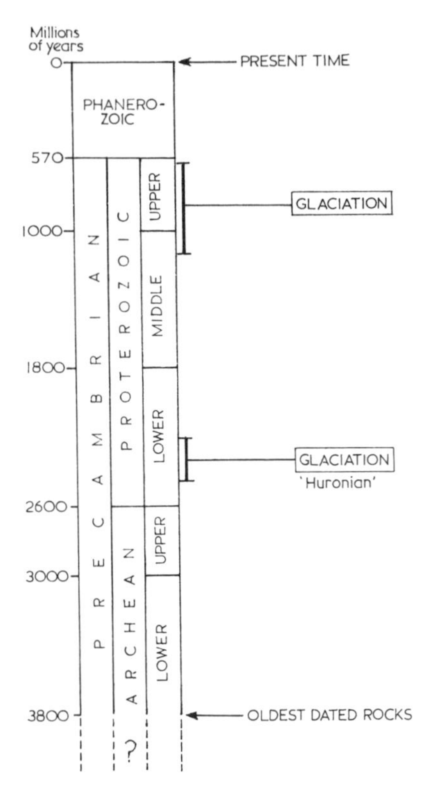

Some major subdivisions of geological time. Note the great length of time involved in the Precambrian. The approximate times of the Precambrian ice ages are also indicated.

Archaean Possibilities

The oldest rock ages to be determined by analyses of isotopes in minerals and rocks are about 3,800 million years, but, as mentioned in Chapter 1, it is commonly inferred from meteorite ages that the Earth is about 4,600 million years old.

A significant part of the early history of the Earth remains shrouded in mystery, for in many regions the oldest rocks are so greatly altered that it is difficult to discover the conditions under which they formed. Most geologists refer to these ancient rocks as Archaean. Some divide the Archaean into two parts, the younger of which contains some areas of well preserved rocks. The younger Archaean includes abundant volcanic rocks which, because of their green colour (mainly due to the mineral chlorite), are called 'greenstones'.

The lavas that constitute Archaean greenstones commonly contain pillows, balloon-shaped masses of rock formed by rapid cooling of an outer skin of molten lava where it was in contact with water. The skin was expanded outward and upward by the pressure of the molten lava below. Such structures occur exclusively in lavas forming underwater, and so are important indicators of the presence of water on Earth in Archaean times. Such 'pillow lavas' are extremely common in the younger part of the Archaean, but are present also in some of the oldest rocks on Earth in the west Greenland region.

Other indications of abundant water at a very early stage in Earth's history include the presence of bedded sedimentary rocks. Such rocks form in two ways: as chemical precipitates from material dissolved in water; and as 'clastic' deposits formed as particles are carried in rivers and streams and then laid down either in river valleys or in larger standing bodies of water such as lakes or the sea. Both chemical and clastic sedimentary rocks from this era show that there was an abundant supply of water, so that the potential for major glaciation probably existed throughout much of Archaean time.

The reasons for the absence of glacial deposits in the most ancient rocks, like the causes of glaciation itself, remain obscure. One possibility is that there was insufficient land surface to support continental glaciers. If the greater part of the world oceans formed early in the history of the Earth and the crust was initially thinner, then most of the crust may have been submerged. In this case any polar 'ice caps' would have consisted of floating sea ice; they could have developed and disappeared without leaving any trace, because they did not erode a land surface. If the Archaean crust were thin (at least locally) and if the interior of the Earth were hotter, then the crust could have undergone much more rapid and pervasive movements due to folding and faulting than is presently the case. These kinds of movements and associated volcanic activity could have had such a profound effect on the sediments deposited that any potential indicators of the ancient climate could have been masked.

In addition to a lack of deposits indicating glaciation, there is little evidence in Archaean rocks of any other specific climatic condition. It has been proposed that the Earth's surface was universally warmer at that time, due, in part, to a higher flow of heat from the radioactive interior. The suggestion has also been made that a denser atmosphere may have played a rôle in making the climate fairly uniform from poles to equator. Again, if the atmosphere contained large amounts of carbon dioxide and hardly any oxygen, as indicated in Chapter 1, this may have inhibited glaciation. These are all speculations; the important fact is the absence of definite evidence of widespread glaciation in Archaean times.

The Advent of the Proterozoic

The transition from Archaean to Proterozoic was, in many parts of the world, a time of profound changes in the nature of the Earth's crust. These changes did not take place everywhere at the same time, but appear to span a range in time from about 3,000 to 2,600 million years ago. They are most spectacularly shown in the sedimentary rocks. Proterozoic sedimentary rocks differ from those of most Archaean successions in showing evidence of stable crustal conditions. One of the attributes of Lower Proterozoic sedimentary successions is the widespread development of thick deposits of slowly accumulated, well washed and winnowed sands that were consolidated into what are known as orthoquartzites. These are described as 'pure' sandstones, because they are white in colour and are composed almost exclusively of the stable mineral quartz (silicon dioxide, SiO_2). Such deposits show that the crust must have remained relatively immobile (gradually subsiding as further sedimentary load arrived) during long periods of geological time. Crustal conditions of this kind were relatively rare during most of the Archaean.

Primeval landscapes, similar to those which may have existed on Earth during Precambrian times: (*above*) a landscape of bare rock and lakes, Norway; (*below*) sandy desert landscape in the Chalbi Desert, North Kenya; (*above right*) a volcanic landscape in the Galapagos; (*below right*) a coastal landscape in an area of columnar basalt, Staffa, Western Scotland.

Pillow lava typical of the greenstones of Archaean age. Such lavas are formed by submarine volcanic eruptions.

The surface of the frozen Arctic Ocean. The earliest ice caps on Planet Earth may have looked like this if there was no land in the polar regions during Archaean times.

Fire and ice. These steep cliffs, in one of the volcanic regions of Iceland, are composed of interbedded layers of ash and ice. Such features may have been common during the earliest glaciations on Planet Earth.

Within the Lower Proterozoic (2,600-1,800 million years ago), significant changes in the Earth's atmosphere have also been proposed. The presence of the minerals pyrite (iron sulphide) and uraninite (uranium oxide) as rounded, transported grains in river deposits near the base of Lower Proterozoic successions in many parts of the world has been interpreted to mean that the atmosphere, at that time, lacked significant amounts of free oxygen (see Chapter 2). The minerals mentioned above are easily oxidized and would not, under present atmospheric conditions, survive transportation of the kind envisaged. They are considered to have formed under unique atmospheric and crustal conditions that were never to be repeated in the Earth's history.

Many geologists believe that most of the oxygen in the atmosphere of the Earth was produced biologically. The major piece of geological evidence for this is the appearance of abundant 'stromatolites' in the Lower Proterozoic. Stromatolites are a special kind of sedimentary rock, produced by the combined effects of organic activity (primitive plants such as algae, together with bacteria) and normal sedimentation processes. The result is a finely layered rock that can exhibit a great variety of forms, ranging from flat layering to club-shaped growths. The fine layering is thought to be due to periods of rapid growth of the cells to produce an organic-rich layer alternating with periods of relative inactivity of the organisms, when the main process was trapping of sediment particles (usually carbonates) by the cellular tissue. In some cases there is evidence that the organisms actually caused precipitation of the carbonate minerals. Most, but not all, stromatolites are preserved in carbonate rocks (limestones and dolostones).

Part of the reason for believing that stromotolites are largely of organic origin is the discovery within them of organic cells. These cells are almost identical to those of blue-green algae which still exist today. Such organisms are photosynthetic—they take in carbon dioxide (CO_2) and produce oxygen (O_2) as part of their life process.

Some stromatolites are known from Archaean rocks, but they first became abundant in the upper part of the Lower Proterozoic (about 2,000 million years ago). The oxygen-producing rôle of stromatolites may be substantiated by the occurrence, at about the same time, of the first extensive red beds.

Red beds are red sedimentary rocks. The red colour is due mainly to the presence of finely disseminated iron oxide (haematite). There are different ideas as to how red beds are formed, but most people agree that they are due to the oxidation of iron-bearing minerals. Some think that reddening occurs in the sediment shortly after deposition; others think that the oxidation took place at an earlier stage, in soils which were later transported and deposited as red sediment. Such large-scale oxidation of iron is taken as evidence that there was free oxygen in the atmosphere. The oxygen may have been derived from the photosynthetic activity of blue-green algae which produced the stromatolites.

Another significant development, at about the same time, was the deposition of very widespread shallow-water iron formations. These are chemically precipitated sedimentary rocks containing large amounts of iron. We will deal with these later.

The Earliest Ice Age

It was on an Earth undergoing significant changes such as the ones we have mentioned that the first glacial deposits accumulated. The Lower Proterozoic rocks of the Lake Huron region in Canada contain some of the best evidence of this ancient glaciation. These sedimentary rocks are about 12,000m thick and include three major layers (formations) that are interpreted as glacial deposits.

Given the thicknesses and the timescale involved, these formations may each represent a different ice age, but they are usually referred together to 'the Huronian glaciation'. The lower two glacial units are not very widespread but the upper unit, the Gowganda Formation, was deposited over an area of at least 120,000 square kilometres and was probably originally much more extensive.

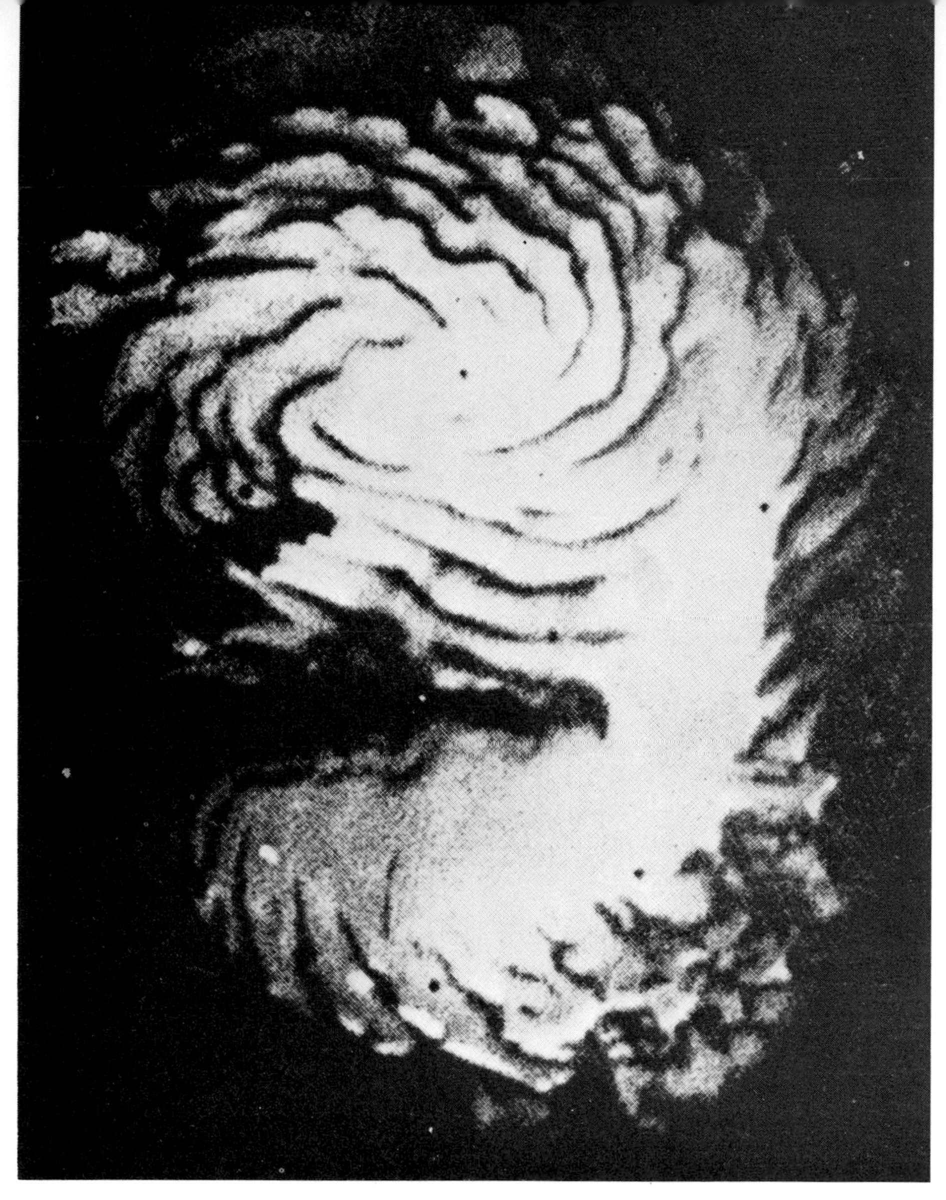

A model of a Precambrian icesheet? Planet Mars, showing the north polar ice cap, possibly formed under conditions somewhat similar to those of the earliest land-based icesheets on Planet Earth.

Evidence for the glacial nature of these rocks is more convincing in the case of the Gowganda Formation than for the two lower glacial units. The most obvious feature of the rocks is their resemblance to tills deposited by Pleistocene glaciers; i.e., they are composed of rock fragments of various shapes and sizes, from various sources, embedded in a fine-grained unlayered matrix that was originally composed of sand, silt and clay particles (see page 77). Using the criteria described in Chapter 3, the rocks appear to be typical tillites, but they were not *necessarily* produced by glacial processes. They may, for example, have formed as slumps or mass flows where matrix and stones slid together down a slope. They may have formed either above or below sea-level. Thus, to prove the glacial origin of such rocks, additional criteria are needed.

As aids to interpretation there are a number of other diagnostic 'glacial' characteristics in the Huronian glacial formations. For example, many of the fine-grained and bedded sedimentary rocks contain dropstones. Also, there are many exposures of rhythmites or varved deposits, which are here interpreted as typical of lakes close to the edges of wasting glaciers. Rhythmites with abundant dropstones are fairly common in the Gowganda Formation. Faceted and striated stones and rock surfaces which have been abraded and grooved by moving ice have also been reported from the Huronian rocks.

Oblique aerial photographs of an ice-marginal lake in East Greenland. Some of the rhythmites found in the Gowganda Formation may well have accumulated in locations such as this.

The Extent of Glaciation

We can say, then, that the Lower Proterozoic rocks of the southern part of the Canadian Shield appear to contain evidence of at least three periods of glacial advance and retreat. The youngest of these, considered to be about 2,300 million years old, left much more extensive deposits than the previous advances. Accepting the glacial origin of these rocks in the Lake Huron region, the obvious question is, 'How extensive was the Early Proterozoic glaciation?'

Deposits similar in age and character to the tillites of the Huronian succession have been reported from several areas of North America. Although fragmentary, the preserved glacial deposits are widespread, extending about 2,000km to the north west and south west of the Huronian occurrence. From detailed studies of the sedimentary rocks, it has been possible to arrive at an estimate of the directions in which they were transported. The main movement of sediment appears to have been radial, suggesting that the glacial sediments were deposited

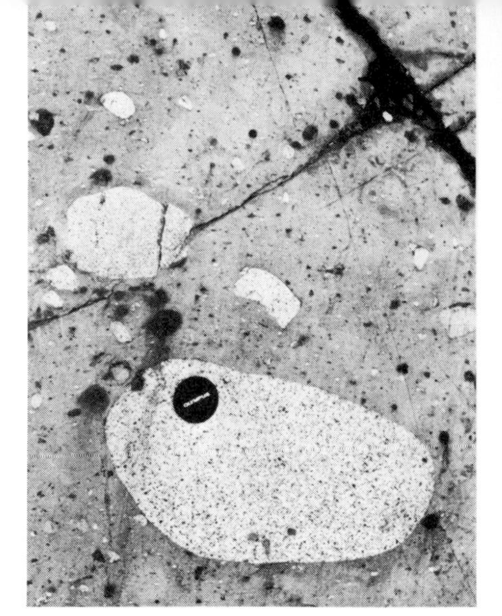

(*Left*) Tillite of the Lower Proterozoic Huronian succession (about 2,300 million years old) on the north shore of Lake Huron, Ontario, Canada. Note the large stones of varied composition, separated by finer grained material.

(*Right*) Finely bedded sedimentary rock formed in a lake near a glacier about 2,300 million years ago. Each thin bed is thought to represent one year's deposition. Note the large stones at the top left. These are 'dropstones' thought to have been carried into the lake by floating icebergs which subsequently melted and released their contained stones.

around the edges of a continental icesheet. The proposed icesheet would have covered a large part of the Canadian Shield, the deposits being preserved only in subsiding basins around its periphery. If this interpretation is correct, and the Early Proterozoic occurrences do in fact represent a high-latitude icesheet in North America, then it is natural to ask if there was another one on the opposite side of the globe—i.e., another polar ice cap.

A search was therefore made for reports of glacial deposits of comparable age on other continents. The best candidate for truly glaciogenic rocks of Early Proterozoic age is a group of rocks known as the 'Transvaal Supergroup' in southeastern Africa. More specifically, the glacial rocks are called the Griquatown Tillite. This tillite forms part of a sequence that is between about 1,950 and 2,340 million years old and lies beneath volcanic rocks that are given a date of 2,200 million years. Another possible glacial deposit of about the same age occurs in northwestern Australia. This supposed tillite is called the Turee Creek Formation and is between about 2,000 and 2,300 million years old —approximately the same age as the South African deposits.

These rocks of North America, South Africa and Western Australia contain the first good evidence of glaciation on our planet. The three regions are presently widely scattered across the surface of the Earth, but their ancient relationships are not known—although the exact positions of the continents in the Lower Proterozoic are presently uncertain, there have been several speculations.

Some clues are obtained from palaeomagnetic studies. From the inclination of magnetism in the Huronian rocks it has been calculated that the proposed ice centre lay at a palaeolatitude of about 60°. In other words, this was the centre of a typical mid-latitude icesheet.

Some reconstructions of the ancient continents place Australia close to

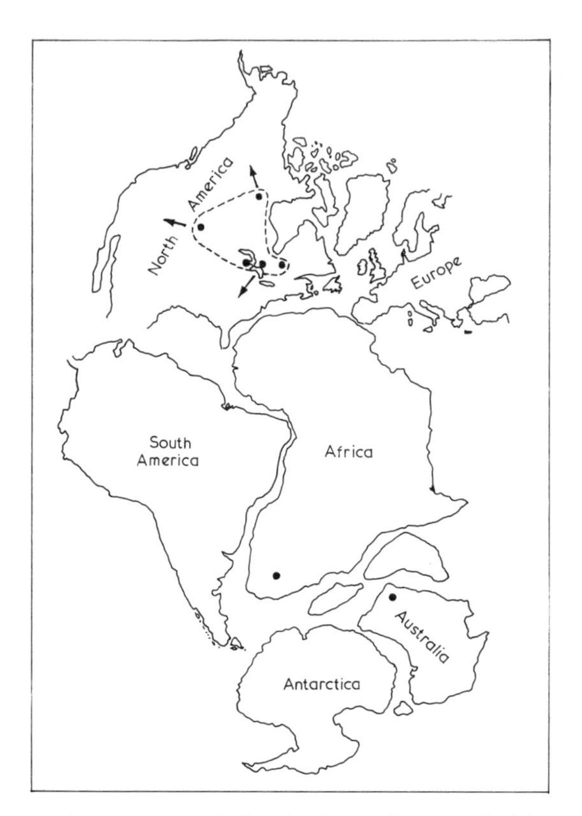

One possible continental arrangement during the Lower Proterozoic (about 2,300 million years ago). Locations where Lower Proterozoic glacial deposits have been found are shown by the black dots. The location of a possible icesheet in North America is shown by the dashed line. Arrows show the radial pattern of sediment transport at that time. Note the presence of glacial deposits of comparable age in southern Africa and western Australia.

North America, whereas in others it is close to South Africa. With such doubts, it is at present impossible to resolve the question of the global distribution of the Lower Proterozoic glacial deposits. One possibility, however, is that there were two major ice centres, one in North America and the other in South Africa/Western Australia, which may have been quite close at that time. Perhaps these two regions represent the ancient high-latitude icesheets of the Early Proterozoic, but at the moment there is not sufficient palaeomagnetic evidence from these rocks to be certain.

In summary, it is reasonably well documented that some parts of the Earth's surface were covered by glaciers as far back as about 2,300 million years ago. It is not certain whether these glaciers represent glaciation of continental type or whether they are related to mountain-forming processes which followed thickening of certain parts of the crust at the end of the Archaean. In any event the appearance of these glaciers followed a period of pronounced change in the Earth's crust, atmosphere and oceans. Perhaps the most important of these changes, in relation to glaciation, was the stabilization, by thickening, of large segments of the continental crust. The continental crust was sufficiently thick, and at a sufficiently high altitude, to support great depths of ice.

An important change in the composition of the atmosphere and hydrosphere

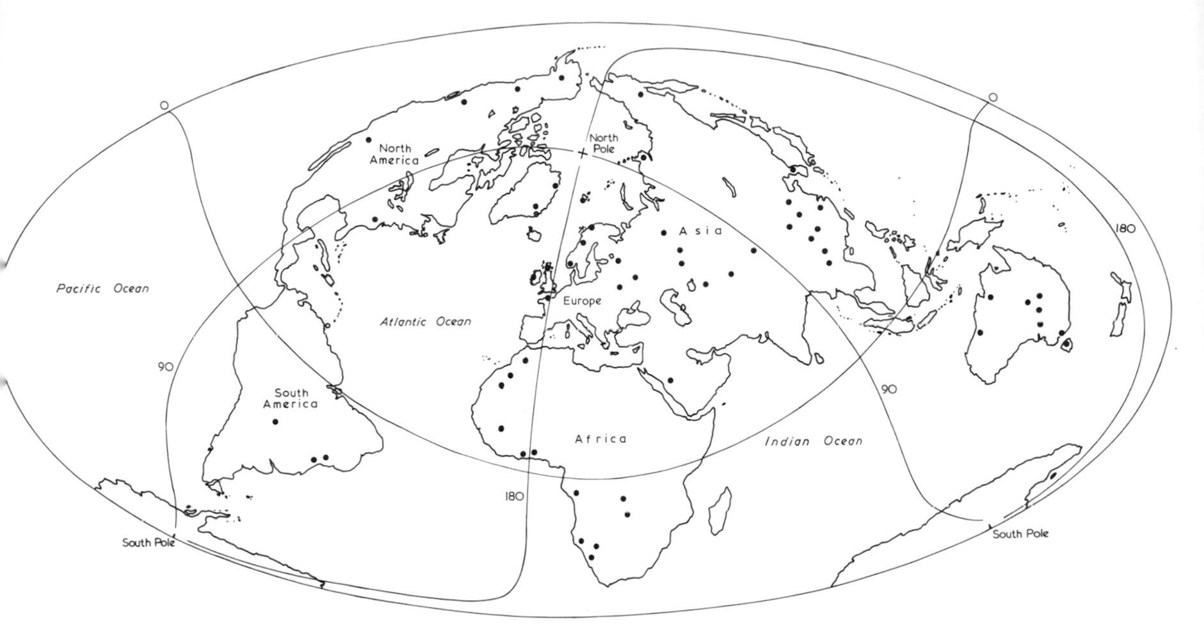

World map to show the widespread distribution of Upper Proterozoic glacial deposits (black dots).

also took place at about the same time as these major 'Huronian' glaciations were occurring. This change, the appearance for the first time of free oxygen in the atmosphere, was probably effected by the photosynthetic blue-green algae, which grew in profusion on the newly established continental platforms and shelves of the Lower Proterozoic. It is witnessed also by the appearance in the geologic record, about 2,000 million years ago, of massive amounts of iron-rich chemical sedimentary rocks. Iron is known to be relatively insoluble when free oxygen is in good supply (witness the lack of dissolved iron in today's oceans and the concentration of residual iron in many present-day soils) but may have been abundant in the oceans of Lower Proterozoic times. The introduction of oxygen into the atmosphere and oceans may have brought about precipitation of iron on a scale never achieved before or since.

Upper Proterozoic Glaciations

During the long period from about 2,300 to 1,000 million years ago there is little or no evidence of glaciation on Earth. The absence of glaciation during this long period is in some ways more difficult to explain than the glaciations themselves, especially in view of the periodicity of ice ages referred to in Chapter 1.

However, the upper part of the Proterozoic is replete with evidence of glaciation. Even if allowance is made for different positioning of the continents, it is difficult to see how all of the glacial deposits could have formed in high and middle latitudes. Several proposals have been made to explain the extent of these deposits, but before dealing with this problem let us take a critical look at some of the evidence for and against Upper Proterozoic glaciation.

Convincing field evidence for Upper Proterozoic glaciation occurs on virtually all continents. The following brief description is limited to evidence from two localities: Scotland and Northwestern Canada. These areas have been selected because of the contrasting environments in which the glacial deposits are thought to have accumulated.

Upper Proterozoic glacial deposits occur in a band that crosses central Scotland in a northeast to southwest direction and continues on the same trend across northwestern Ireland. These rocks belong to the Dalradian succession, the uppermost part of which has yielded fossils considered to be of Lower Cambrian age. Near the western extremity of the outcrops in Scotland, on the island of Islay and particularly on some small islands known as the Garvellachs, there are spectacularly well exposed and preserved examples of the Dalradian glacial deposits. These rocks are known as the Port Askaig Tillite. They were the subject of an extremely detailed account by A. M. Spencer, who documented several lines of field evidence in support of a glacial origin.

The Port Askaig Tillite is up to about 870m in thickness and contains 47 'mixtite' units interpreted as tillites. The formation as a whole is bounded above and below by carbonate rocks. The individual mixtites are separated from each other by siltstones, sandstones, conglomerates and dolostones, all probably deposited under water. The field characteristics described by Spencer as indicating a glacial origin for the Port Askaig Formation include its great thickness and area of extent. He also cited the widespread occurrence in the mixtites of erratic stones which had been derived from outside the depositional basin and had therefore been carried a long distance (presumably by ice). In addition, Spencer reported the presence of large polygonal cracks on the upper surfaces of some mixtite units. These cracks are filled with sandstone and conglomerate; they were considered to be ancient examples of the ice wedges that are present today in periglacial regions. The combination of the two features—rhythmically bedded fine-grained sedimentary rocks, and the presence of large isolated stones—was considered by Spencer to be the most definitive indication of glacial origin.

Prior to his work, most people interpreted the Port Askaig and similar tillites as deposits from floating ice. Spencer suggested that, in fact, the extensive mixtites were the product of deposition from grounded icesheets on a stable marine platform. As supporting evidence for his ideas he cited the presence of sharp basal contacts (floating ice would have resulted in an interbedded contact) to the mixtites, and the shallow water origin of interbedded sediments. He explained the presence of discontinuous bedding in the otherwise massive mixtites as being due to meltwater streams at the base of the icesheet.

Based on the evidence from these islands of western Scotland, Spencer developed an interpretive model that involved the advance and retreat of icesheets over a broad shallow marine shelf. This model explains the presence of extensive mixtites, interbedded with sandstones and other layered sedimentary rocks, all formed on a shallow shelf sea environment. Following melting of the grounded icesheet and the deposition of till, the ice receded and permafrost conditions produced the ice-wedge cracks that were later filled with sediment and preserved on the tops of many mixtite units. Rises in sea-level, following melting of the glaciers, caused deposition of bedded marine sediment once more. In this way Spencer interpreted the Port Askaig Tillite in terms of seventeen

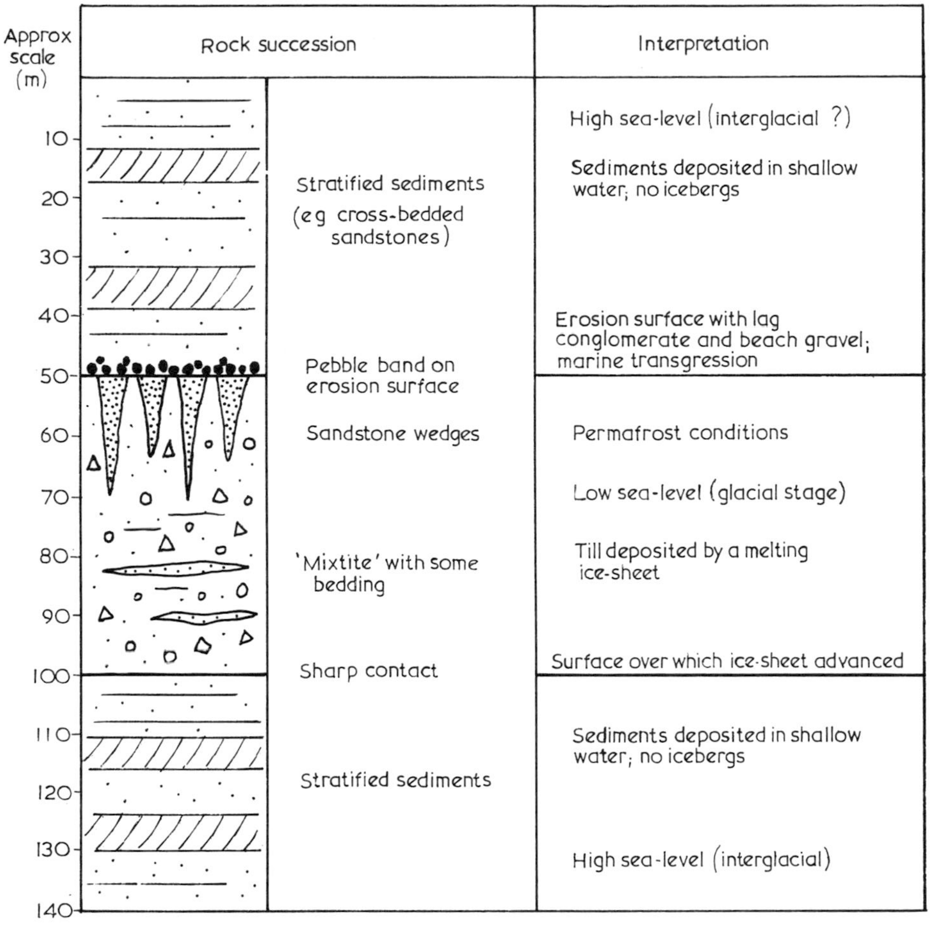

The sequence of sediments typical of the glacial-marine cycles in the Port Askaig Tillite. Each cycle can be interpreted as a glacial-interglacial cycle, and the tillite contains at least seventeen such cycles.

major advances of the ice, indicating an ice age of considerable duration and complexity.

In contrast to those of the Dalradian sequence of Scotland, Upper Proterozoic glacial deposits of many regions were deposited and preserved in areas where there is good evidence of contemporaneous faulting, suggested by some to be an indication of the onset of continental break-up. The western part of the North American continent contains several examples of this kind of deposition. The following brief description concerns one such area near the north end of the Cordilleran fold belt.

The tectonic setting appears to be in marked contrast to that inferred by Spencer in Scotland for the deposition of Upper Proterozoic glacial sediments. The North American rocks were probably deposited in an elongated trough that shows clear evidence of faulting before and during the period in which they accumulated. The glacial rocks in question form part of the Rapitan Group. The lowest unit is composed mainly of purple mudstone and purple-to-grey sandstones together with some conglomerate layers. There is good evidence

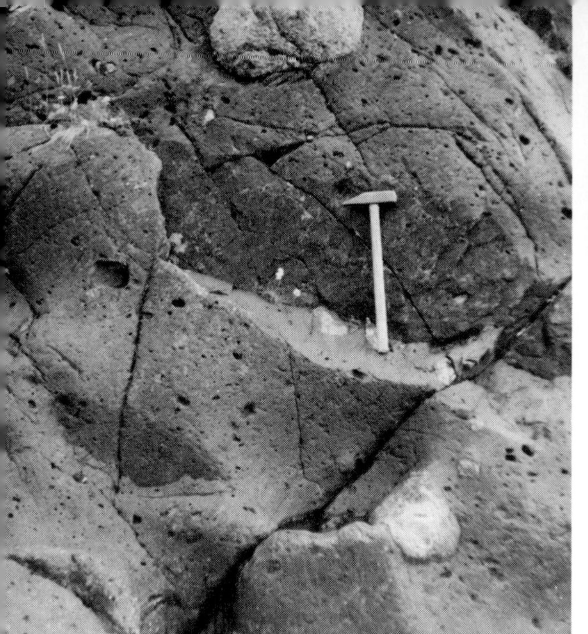

(*Left*) One of the mixtites from the Port Askaig tillite sequence, Garvellachs, Scotland. The deposit contains granite erratic boulders and also dolomite stones, some of which have weathered away to leave holes in the rock surface.
(*Centre*) An unbedded mixtite at Port Askaig containing an abundance of erratics of dolomite and granite.
(*Right*) Varved siltstones in the Port Askaig sequence. These waterlain deposits contain ice-rafted 'dropstones' of various types.
(*Below right*) Sandstone wedges cutting the top surface of one of the mixtite layers at Port Askaig.

that some of these stratified sedimentary rocks were brought into a subsiding basin by means of turbidity currents and related mechanisms. Such deposits indicate episodic downhill movement of sediment, triggered off by, among other things, contemporaneous earthquakes and fault movements. These generally finely bedded sedimentary rocks include some large isolated stones, some of which appear to have been dropped in. The simplest explanation is that they were carried into the depositional basin by icebergs or floating shelf ice, and this suggestion is confirmed by the presence of stones that carry striations. The purple colour of these rocks is due to the presence of finely disseminated haematite (iron oxide), and in some localities an iron formation occurs at the top: this iron formation also contains dropstones and must have formed during the glaciation.

The succeeding unit of the Rapitan Group is a thick purple, buff and green sequence containing mixtites. In many places there are interbedded siltstones and sandstones that contain structures suggesting the activity of water currents. Therefore it is probable that the conglomerates and mixtites were formed in a marine environment. The presence of several distinct mixtite units might indicate that there were several advances of an ice-shelf or floating icesheet out over the marine basin. The depth of the water is, however, unknown. Some highly contorted sedimentary layers suggest slumping on an unstable slope. Glacial influence is shown by the presence of scratched and striated pebbles.

Above the mixtite unit there is a thick, dark grey shale with some silts and sands interbedded. There are some indications that turbidity currents played a rôle in the deposition of this unit. There is no direct evidence of glacial

influence on the deposition of the black shales, but a deepening of the basin may be reasonably inferred from the nature of the sediments. This may reflect a rise in sea-level following melting of the glaciers that were present during the deposition of the two underlying units.

These two examples serve to illustrate some of the field evidence for glaciation in the Upper Proterozoic and to emphasize the variability of the record that it left. An even greater spectrum of depositional environments could be illustrated by inclusion of truly continental glacial deposits such as those of the Upper Proterozoic of northwestern Africa.

The Scottish example is interpreted as an example of Upper Proterozoic glaciation on a stable marine shelf, whereas the example from western North America typifies marine glacial deposits of a fault-bounded, subsiding marine basin. It is probable that the geological record of ancient glaciation has a strong bias towards marine deposits of rapidly subsiding basins because such sediments have a much greater chance of survival than those of continental or even shallow marine environments.

Problems of Interpretation

Although most geologists who have worked on Upper Proterozoic sediments accept the evidence for widespread glaciation, some authorities, such as L. J. G. Schermerhorn, have argued forcefully that many so-called 'glacial' features are best interpreted in other ways. There has been much discussion of the real importance of features such as tillites, dropstones, till pellets, striated erratics, and abraded rock surfaces.

Tillites have already been discussed in the previous chapter, and we have mentioned the various way in which 'apparent' tillites can be produced. With or ice ages, L. J. G. Schermerhorn has suggested that there are few genuine Upper Proterozoic tillites. He suggested that, in some cases, earthquakes may cause material, including large blocks, to move down slopes. Such blocks may even become scratched by rubbing against each other as they move, thus resembling stones deposited from glaciers. Another way in which glacial-looking deposits can be formed involves sliding and slumping of sediments down an underwater slope. Under some circumstances, large stones may be present on such slopes and, if incorporated into the slump deposit, may form a till-like sediment. Where an earth-tremor or some other disturbance causes sediment to be raised as a cloud into the surrounding water, the sediment-laden water, because it is denser, will tend to move down even a gentle slope as a turbidity current. The material deposited by such a current is commonly graded from coarse at the base to fine at the top because the larger particles settle out of the deposit before the finer ones. Such deposits may also contain stones and may superficially resemble some kinds of glacial deposits.

There has also been discussion concerning the origin of dropstones in sedimentary rocks of Precambrian age. Normally they are interpreted as the result of ice-rafting, but even stones dropped from floating ice do not necessarily indicate that there were *glaciers* present, for ice developed by freezing along the shore of a lake or sea may incorporate stones into its base. When such ice is fragmented and dispersed by winds and tides the contained stones may melt out and drop into bedded sediments below. In younger rocks, formed after the evolution of larger animals and plants, such stones might conceivably be transported and dropped by animals or from tree roots or other plant materials. Another possibility is that such stones could have been thrown from erupting volcanoes. In this case, the stones would mostly be volcanic rocks. None of these possibilities explains the great abundance of dropstones of varied composition in Precambrian rocks, and one is forced to the conclusion that glacial ice was in fact present.

Some Precambrian deposits contain till pellets, which are like dropstones but generally smaller. Usually they are composed of fine-grained silty sediment interpreted as till. They can form in or adjacent to glaciers in a number of different ways, especially where tills are 'washed' and eroded by meltwater or subjected to wave attack on a beach. Till pellets have been described from icebergs off the coast of Alaska, and they are generally considered to be good evidence of glacier ice. They have been reported from the Lower Proterozoic Gowganda Formation and from the Upper Proterozoic glacial deposits of the northwestern part of Canada. However, they are reliable indicators of glaciation only if they are *genuinely* composed of till.

Another feature that is difficult to explain by a non-glacial hypothesis is the presence of the striated and grooved erosion surfaces on which some supposed Upper Proterozoic glacial deposits are known to lie. Such surfaces have been reported from Upper Proterozoic rocks in several parts of the world. It has been suggested that scratched surfaces could be developed under mudflows, but such occurrences are not liable to be very extensive; glacial erosion provides by far the most convincing explanation for the majority of striated rock slabs.

Striated stones are found also in many of the deposits referred to above. Again, a number of non-glacial processes have been invoked in order to explain them. Some scratched stones may be produced by movement of stones during the deposition of mudflows. In folded strata, the scratching may have formed much later than the time of deposition of the sediment; folding of rocks may cause movements between stones and their matrix, and such movements may, in exceptional circumstances, leave scratches on the surfaces of the stones.

Glacially striated stones are, however, fairly easy to recognize. The scratches are commonly developed on a flattened or abraded surface, and they may run in several cross-cutting directions. It is especially significant if small, hard stones are striated, for these are less likely to occur in non-glacial sediments such as mudflows.

There has also been some discussion among geologists concerning the significance of polygonal markings. These occur on the upper surfaces of some of the mixtite units of the Port Askaig Formation in Scotland, and also on the Upper Proterozoic 'ice age' rocks of northwest Africa. Normally, as indicated in the previous chapter, such markings are attributed to the contraction of the ground due to extreme cold or permafrost conditions. This interpretation is reinforced if ice-wedge 'casts' can be identified. On the other hand, it has been argued that patterned ground can be created by desiccation, as when the bottom of a pool dries out, and that wedges can also be created by the concentration of salts in hot desert areas. But the arguments in favour of permafrost conditions must be conclusive if the patterned ground is widespread and also associated with tillites and other indicators of cold climatic conditions.

Striated bedrock of Upper Proterozoic age, in southern Sweden. The glaciated rock surface is overlain by Cambrian sediments.

One problem that has been noted by a number of geologists is the close association between dolostone (a rock composed mainly of the carbonate mineral dolomite) and supposed glacial deposits. In some cases beds of dolostone lie directly on top of supposed tillites. This poses a problem, because dolostone is presently known to form only under relatively warm conditions such as in the Persian Gulf and in Western Australia, whereas tillites presumably formed under cold conditions. In some cases the dolostones have fine bedding resembling varves and may even contain dropstones. The formation of dolostone is very complex and not fully understood. One possible explanation for some of these occurrences is that the dolostones formed by sedimentation of small dolomite particles eroded by the glacier from older rocks made up of dolostone. The dolostones associated with the tillites would thus have formed by a process of resedimentation under cold conditions, although the original dolostone, from which the particles were derived, could have formed in a hot climate.

Alternatively, if some of the dolostones associated with glacial deposits are in fact 'primary' deposits, then we must either call for very special conditions (a difference in the composition of the atmosphere?) for formation of the dolostones under cold climatic conditions, or refute the glacial origin of the associated rocks.

Similar arguments have been applied to the presence of iron-rich sedimentary rocks associated with supposed Upper Proterozoic glacial deposits in various parts of the world. These iron formations are a headache to proponents of the idea that there was a changeover in the atmosphere from one containing virtually no free oxygen to one containing at least some free oxygen, about 2,000 million

Stone removed from an Upper Proterozoic tillite in northwestern Canada. Its surface is covered with glacial striae or scratches.

years ago. According to this idea, the abundant iron of the ancient ocean should have been precipitated at about that time, never to be replenished because of the insoluble nature of iron when oxygen is plentiful. How then can we explain the iron formations that are often intimately associated with Upper Proterozoic ice ages? It has been argued that such iron-rich sedimentary rocks must have formed under tropical conditions, for red, iron-rich sedimentary soils (laterites) form today in such regions. The Upper Proterozoic iron formations are, however, not soils, but appear to be chemical precipitates, probably formed in a marine environment. Even if the iron was derived as insoluble material concentrated in soils under oxidizing conditions, what was the mechanism of dissolving, transporting and precipitating it? It has been suggested that the answers to some of these questions may lie on the ocean floor.

It is well known that iron is a common constituent of the kinds of lavas and other igneous rocks that form on the sea floor. Such rocks are especially abundant and well exposed at mid-ocean ridges, the great submarine mountain chains that as we saw in Chapter 1 are the sites of formation of new crustal material from the underlying mantle. The rocks of the mid-ocean ridges are at higher temperatures than those of the surrounding sea floor and are thought to be regions where vast amounts of sea water circulate through basic igneous rocks. Recent experiments, and observations at mid-ocean ridges, suggest that, under such conditions, chemicals such as iron and silica are soluble in the heated sea water. These substances tend to precipitate when the heated, iron-rich sea water is diluted with normal sea water.

Such a mechanism has been suggested to explain the iron formations associated with many Upper Proterozoic glacial deposits. The rôle of the glacial ice is seen to be as follows. When a glacier reaches sea-level, it tends to thin and spread out because of reduced bottom friction. Thus a floating ice shelf can form. Cold, dense waters from the base of the ice shelf would flow into the deeper parts of the basin, where they could displace iron- and silica-rich waters, causing them to rise and flow shorewards. Dilution with the cold, sub-glacial waters could then cause precipitation of the iron and silica to produce an iron formation. Such iron formations are envisaged as forming in an environment quite close to an oceanic ridge system. This suggests a small ocean basin, perhaps comparable in size to Baffin Bay, with glaciers descending into it.

Extent of the Glaciations

Assuming that the sum of evidence available supports the concept of glaciation in the Upper Proterozoic, let us now return to the problem of explaining why it was so widespread.

The first idea was that the Earth underwent a very extensive single ice age with icesheets extending even into tropical latitudes. This idea is quite exciting because it means that the glacial deposits, formed at approximately the same time all over the world, might be used to correlate Precambrian rocks. In Phanerozoic rocks, which are younger than the beginning of the Cambrian, most correlations are based on fossils, the assumption being made that a similar assemblage of fossils means that the rocks are of approximately the same age. On the other hand the correlation of strata is particularly difficult in the Precambrian because there are few if any fossils.

As more and more Upper Proterozoic glacial deposits are dated, it is being found that they are by no means all of the same age. In fact, they are now known to range in age from over 1,000 to about 570 million years ago, the date normally taken as the beginning of the Cambrian. As described in Chapter 1, listing of all the known dates suggests that most of the glacial deposits cluster around 600 million years old, but others cluster around 750 million years and yet others are about 900-950 million years old. Thus the concept of a single widespread and geologically instantaneous ice age that can be used as a 'time marker' for Upper Proterozoic rocks has to be abandoned. It is much more likely that there were *three* Upper Proterozoic ice ages. As is commonly the case in geology, the story is more complex than it first appeared in our foreshortened view of geological history.

It was mentioned in Chapter 2 that the Upper Proterozoic glaciations may have been triggered off by what has been called the 'anti-greenhouse effect'; i.e., the supposed influence of loss of CO_2 from the atmosphere just before the onset of glaciation. This decrease in atmospheric CO_2 was postulated because, in many regions, the glacial deposits lie on top of thick sequences of rocks made up of carbonate (limestones and dolostones). Formation of such carbonate rocks would have involved withdrawal of significant amounts of CO_2 from the atmosphere; this would have allowed more reflected heat energy to escape from the Earth's surface, and hence have caused an overall drop in temperature. This idea is, however, not without its problems, because there are also abundant carbonate rocks *above* the glacial deposits, and there are many examples in the geological record of widespread deposition of carbonate rocks without attendant glaciation.

Another possible mechanism has been put forward to explain the widespread distribution of Upper Proterozoic glacial deposits, and the fact that some of them appear to have formed in tropical latitudes. This idea invokes variation in the orientation of the spin axis of the Earth in relation to the plane in which the planets rotate around the Sun. It has been calculated that if the Earth's rotational axis tilted at an angle greater than 54° then the amount of energy received from the Sun, during a year, would be greater at high than at low latitudes. Thus, if the tilt angle were gradually to increase from 0° to 90°, then, given the conditions suitable for the onset of glaciation (say, a decrease in the overall energy output of the Sun) glacial conditions should migrate from high to low latitudes. This elegant, but speculative, hypothesis would explain the apparent low latitude of some glacial deposits, but, as stated above, much of the palaeomagnetic evidence for low-latitude glaciation is suspect.

Widespread glaciation in the Upper Proterozoic has also been explained as the result of a period of rapid continental drift, during which the continents successively moved through polar latitudes and were glaciated. According to this idea, the glacial deposits should be of different ages on different continents, and therefore not useful for dating and correlating the rock sequences containing them. Choice between this idea, and the concept of extremely widespread contemporaneous glaciation of most of the globe, would be easier if good palaeomagnetic data were available from the tillites in question. As pointed out in Chapter 2, a steep inclination of the 'locked in' magnetism in the rocks should indicate deposition in high palaeolatitudes, whereas a near-horizontal magnetization should indicate deposition near the equator. Initial studies of this kind were said to support the concept of glaciation in low latitudes, but subsequent work, in some cases at least, has shown that the magnetic history of these old rocks is more complex than was initially anticipated. Some such rocks have been found to contain several magnetic directions and some of those initially thought to contain evidence of glaciation in tropical regions (i.e., a very widespread contemporaneous glaciation) are now considered to have formed in high palaeolatitudes.

The most viable alternative to worldwide glaciation is the idea that the so-called glacial deposits of the Upper Proterozoic were controlled by a special kind of mountain-building process related to the fragmentation of the earliest continents; this may have marked the onset of plate tectonics. Recently, much effort has been expended in attempts to discover whether the concepts of plate tectonics and sea-floor spreading can be applied to rocks which are more than 570 million years old. Some people have suggested that plate-tectonic processes, involving the opening and closure of ancient oceans, have been going on since Archaean times, more than 2,600 million years ago. Other geologists think that the process of continental break-up began slightly later, while yet others suggest that such fragmentation was initiated about 800-900 million years ago. The break-up of early continental masses, probably initiated above particularly hot parts of the mantle, was probably accompanied by uplift of the edges of the newly formed ocean basins. This kind of setting is ideal for the development of mass flow deposits which might look much like glacial deposits.

The story might be further complicated by the formation of mountain glaciers on the elevated rims of the ocean basins: at intermediate to high latitudes such glaciers could have descended to the sea; and their deposits would have been preserved in the small ocean basins, together with the slump deposits and turbidites related to earth-tremors and fault movements during uplift of the basin margins and subsidence of the basin floors.

As pointed out by L. J. G. Schermerhorn, many of the so-called glacial deposits of the Upper Proterozoic are preserved on the margins of such fault-controlled basins. According to this idea, the most important factor was mountain-building activity rather than the onset of a severe climatic régime (see page 35). There is in many regions, however, unequivocal evidence that glacial deposition, perhaps accompanied by slumping and mass flows, did take place in the Upper Proterozoic on a large scale. Although mountain glaciers may have been responsible for these deposits in some areas, there can be little doubt that there was also widespread icesheet glaciation during each of the Upper Proterozoic ice ages.

Choice among the various ideas concerning the nature and origin of the glacial deposits of the Upper Proterozoic is still open. The early ideas of a single continental glaciation of global extent now seem to be discredited. As additional data from palaeomagnetism and radiometric dating become available, it is probable that more and more evidence will emerge of glaciation during a series of at least three separate ice ages. The initiation of these ice ages probably involved both vertical movements of the crust (to allow mountain glaciation) and horizontal movements of the continents (to allow different continents to move sequentially into high and middle latitudes). Thus we have an exceptionally widespread record of glaciation in the Upper Proterozoic. The ice ages of the Upper Proterozoic may prove to have been favoured by special mountain-building episodes combined with rapid continental drift; but the root cause of each period of ice expansion is most likely to have been global climatic cooling.

GRANT YOUNG

5

Traces from the Desert: Ordovician

The existence of an Ordovician ice age has been confirmed only within the last decade, largely because most of the evidence for it comes from remote locations in the Tropics where geological work is still at a reconnaissance stage. It was, however, an ice age of considerable duration (possibly 25 million years or so) which affected a vast area of the old supercontinent of Gondwanaland. As during the ice ages of Precambrian time, the Ordovician icesheets appear to have been partly continental in character and partly maritime, and there is evidence that at least one of the icesheet margins was a zone of great glaciological instability. Field information is still being accumulated rapidly, and it is certain that within the next decade our sketchy impressions of this ice age will be clarified and perhaps substantially altered.

The Ordovician world was very different from that of the present day. As mentioned in Chapter 1, the Ordovician period, which lasted from about 500 to about 440 million years ago, was a time when the continental landmasses were quite widely dispersed, with a number of different cratons (Precambrian shield areas) whose outlines must have looked vastly different from the main continental landmasses of today. However, we know that the ancient cratons of North America, Europe and Siberia were located in tropical or subtropical areas, while part of the supercontinent of Gondwanaland was also in the tropics. There appears to have been no land near the north pole, but in the southern hemisphere a large part of Gondwanaland was located in high latitudes. Much of the area above 60°S was dry land, and the south pole was displaced by continental drift from what is today northwest Africa in Cambrian time, across the Saharan region, and into the vicinity of the Congo Basin by the late Ordovician. During Silurian time the supercontinent continued to drift across the south polar regions.

Large parts of the continental margins were submerged by shallow seas during the Ordovician period, and there were a number of distinct marine transgressions and regressions. Many of the sedimentary rocks are of marine origin—fossil-rich limestones, sandstones, siltstones and shales. There are also conglomerates and grits. In deeper waters, rocks such as mudstones, silty limestones and black shales were laid down. Sea floor spreading was quite active during the Ordovician, and there was a gradual evolution of geosynclines on the flanks of the ancient cratons. Collisions between plate margins created a number of mountain belts, and in places there was a great deal of volcanic activity.

Overall, there must have been rapid erosion of the land surface during the

Ordovician, for there was no land vegetation to ameliorate the effects of the runoff of surface water. Probably there was rapid weathering of exposed rock surfaces and rapid removal of weathered materials and unconsolidated sediments by myriads of streams and rivers.

In some areas, such as the continental interiors of 'Ancestral Siberia' and in the Australian part of Gondwanaland, the climate was warm and dry, leading to the formation of red beds and evaporites. In high latitudes, on the other hand, it is quite likely that permafrost existed in the continental interior during the period of climatic cooling which culminated in the Ordovician ice age. In the middle- and high-latitude seas, water temperatures fell below the tolerance limits of most of the organisms called graptolites, which had previously been particularly numerous in these areas. By about 480 million years ago world climate had cooled substantially, and the Ordovician ice age was under way.

Convergence of Evidence

The actual discovery of the Ordovician ice age is one of the most intriguing stories in the recent history of Earth science. It involves a convergence of evidence from three completely disparate approaches.

The first concerns the long-term cyclic pattern of geological events. The topic has already been discussed in Chapter 2, and we have also discussed the apparently regular spacing of ice ages through geological time. In view of this regularity of ice ages, and in view of the fact that the winters of the world seem to occupy no more than about 10% of total geological time, it seemed curious in the 1950s that no ice-age effects had been found midway between the Eocambrian (around 650-700 million years ago) and the Permo-Carboniferous (around 250-300 million years ago). I sketched an idealized cycle and highlighted this apparent omission in 1960; independently and unknown to me, a Soviet scientist named Lungershausen made the same proposal at about the same time, or even earlier.

The second line of approach has been by means of palaeomagnetic studies (see page 178), notably those of the Australian geophysicist Michael McElhinny. By studying the remnant magnetism preserved in suitable types of rock and working back through time, the South Pole's position has been traced back from its present site in Antarctica, where it has been since Permian time, through South Africa and Brazil, to, by Ordovician times, North Africa. I suggested that this 'Gondwana Pole' should be the site of glacial formations.

Now comes the third and crucial step in this history of discovery. A number of scattered surveys in West Africa and the Sahara by French petroleum geologists, over the period 1960-1970, had been turning up curious evidence of giant boulder beds where some of the boulders looked as if they had glacial scratch marks. Then, in southern Algeria, east of the Hoggar Mountains, Serge Beuf and some associates from the Institut Francais de Pétrole stumbled across a long series of giant grooves that had been uncovered by the drifting sand dunes of the central Sahara. Convinced they were glacial, the Institut, jointly with Algerian geologists, organized an international team of glacial experts to come to view the evidence. For several weeks we were trekking across the Sahara in land rivers, climbing hills and carrying out low-level reconnaissance flights. Finally, we came to the giant grooves. They were found to be

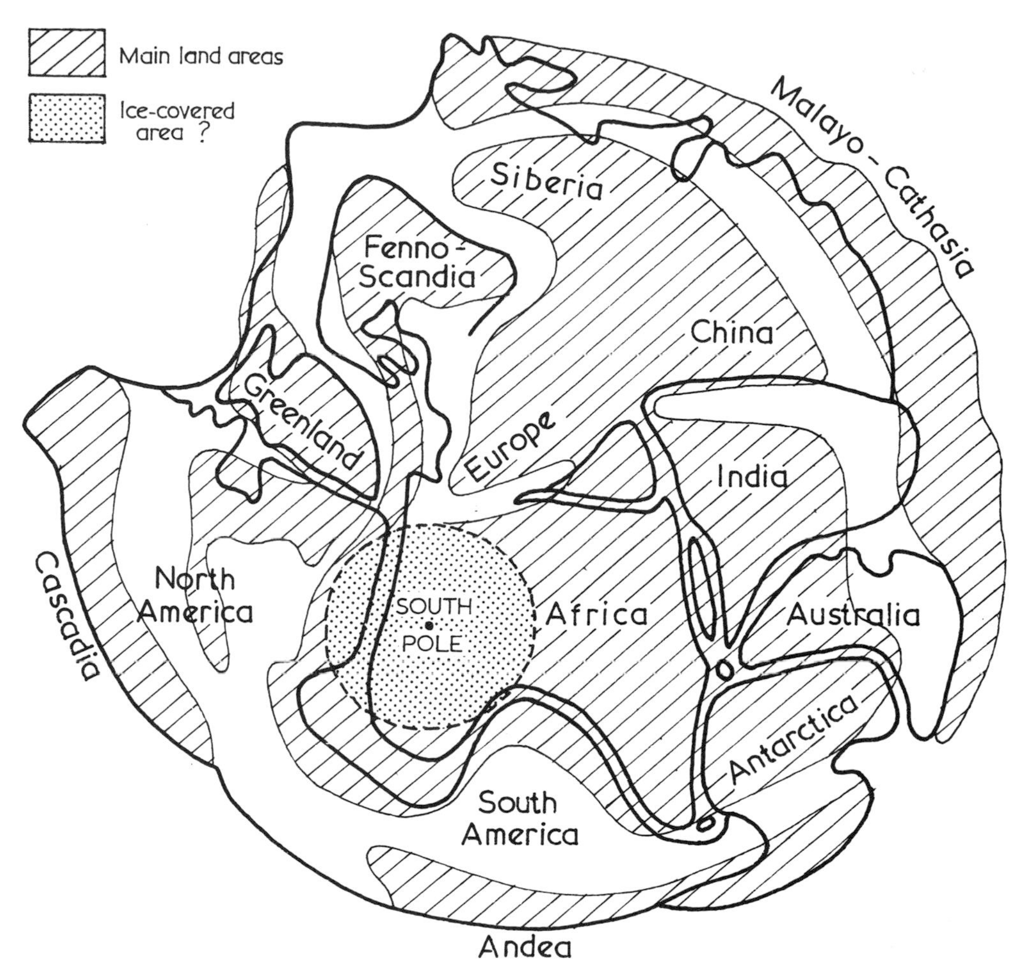

A little known reconstruction of Ordovician Pangaea by A. W. Grabau, published in China in 1940. It is of great scientific interest that Grabau was able to present this palaeogeographic model on the basis of studies of fossils and stratigraphy, which showed clearly that West Africa was then a cold region and that the southern continents were once all connected by land. Grabau could not possibly have known about palaeomagnetics or plate tectonics theory, which only appeared decades later.

marked with little half-moon-shaped scars, typical of those created by the tremendous friction generated by rolled pebbles at the base of modern glaciers; there were scour-shaped bowls, of the sort created by powerful turbulence in subglacial streams; there were boulder-beds of exotic rock types, transported from hundreds of kilometres away; and there were vast formations of yellow quartz sands typical of the outwash of melting glaciers as, for example, in Iceland today. And all these key bits of evidence lay within a fossil-bearing formation of Upper Ordovician age. In this formation there were trilobites, brachiopods and curious vertical tubes known as *Skolithos*, the entire collection being extraordinarily similar to the well known fossil beds of this age in North Wales.

For the visiting scientists the effect was electrifying. Here we were privileged, beneath the hot Sahara sun, to see the detailed record of a giant glaciation, precisely dated, and just where it had been predicted to be by the evidence of the palaeomagnetists. Our French hosts were not unprepared for the occasion.

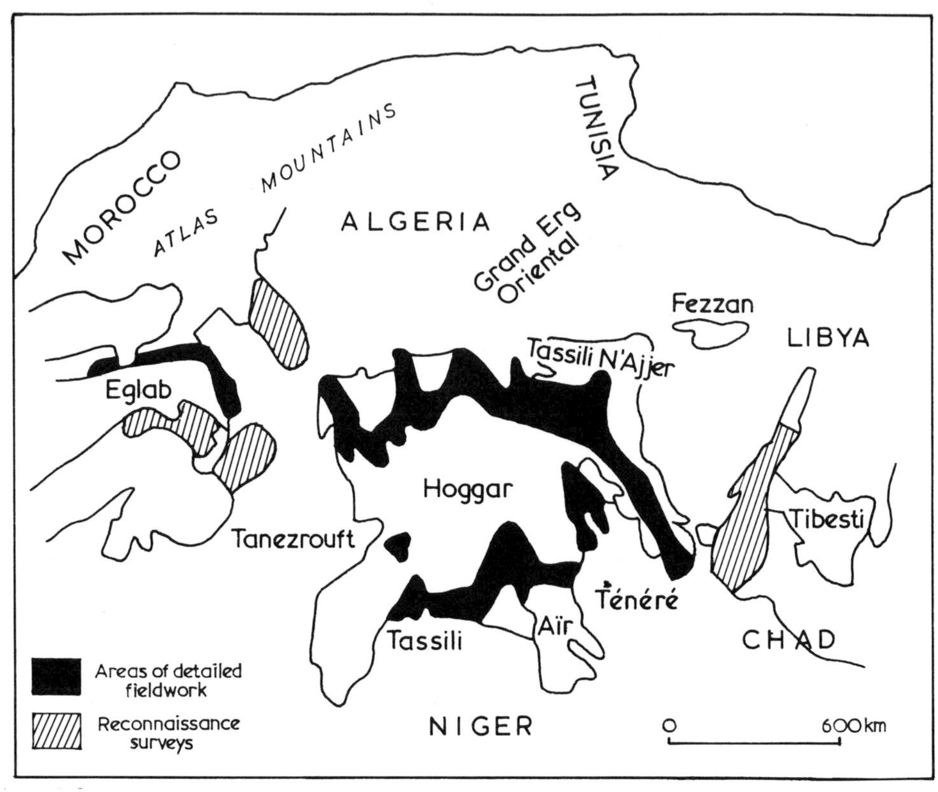

Map of northwest Africa, showing the areas where French petroleum geologists made the main discoveries of Ordovician glacial features in the period 1960-70.

There was a refrigerator on our supply vehicle, and out of it miraculously emerged a bottle of the finest champagne, ice-cold. And so we drank to the health of the discoverers, to the visitors, and to the Ordovician ice age!

Distinctive Features

By 1970 the basic conclusion had been reached that here indeed, in the midst of the world's hottest desert, at around 25° north of the equator, was unquestioned evidence of a major continental glaciation. Now geologists began to ask themselves about the meaning of the unusual or distinctive features of the Ordovician ice age. They also began to ask about the glaciological characteristics of the Ordovician ice, and about the extent and duration of the glaciation.

First of all, perhaps the most striking thing is that the key evidence comes from the middle of an immense Precambrian shield, more than 1000km—in any direction—from any ancient or modern plate margin. The underlying 'basement' of Precambrian gneisses, schists and granite is a smoothly eroded ancient peneplain, unconformably capped by almost flat-lying Ordovician sedimentary rocks which are mainly sandstones. The unconformity is often marked by a band up to a few centimetres thick of brilliant red haematite. Could this be a relic of a Precambrian or Cambrian soil? At the present day, red iron-rich soils are strictly limited to the moist tropical zone.

Following for long distances across the West African shield, one sees that the

Part of the Upper Ordovician glacial sequence of the Sahara Desert. In the middle of the sequence is an unstratified tillite containing white erratic blocks. The cross-bedded formation below is a continental shelf sandstone, whereas the curiously jointed massive formation above is an old outwash sand, left by the melting glaciers.

A characteristic outcrop of sandy Ordovician tillite formations in the central Sahara, east of the Hoggar and south of Djanet in Algeria.

Linear striations and grooves left on the underlying sandstones by the Ordovician ice motion, which was from south to north in southern Algeria. Similar grooves can be followed for hundreds of kilometres, except where they are covered by desert sands.

Crescentic patterns created by powerful meltwaters beneath the ice (under strong hydraulic pressure). Similar features are found in Sweden in some areas affected by the Scandinavian icesheet of the last glaciation.

overlying Ordovician sandstones are not completely flat-lying. They have been warped into broad domes and basins, with a 'wavelength' of about 500-1000km and an amplitude, from crest to basin, of 3,000-5,000m. The dip of the formations is thus usually almost imperceptible at any particular point, although in places there has been some comparatively recent faulting, and here the beds may be locally tilted up to steep angles. These fault lines are mostly revivals of the old Precambrian lineament trends, dominantly north-south in this part of Africa; they show up very distinctly on the satellite photographs in this open country, bare of vegetation or soil. Some of the faults must have been reactivated only within the last million years or so, for they have provided pipes along which molten rock from the mantle has travelled towards basalt volcanic eruptions. The most impressive of these volcanic centres is in the Tibesti dome, but smaller clusters are seen in the Hoggar, Aïr and elsewhere.

The flat nature of the pre-Ordovician peneplain is of critical importance in the discussion of the exact nature of the Ordovician glaciers. Some writers have said, to paraphrase their remarks; 'There are plenty of high-mountain glaciers today in tropical latitudes. Your African evidence simply shows that the Hoggar was uplifted at that time.' But the peneplain surface passes over the Hoggar with only a gentle up-arching. Had there been a high mountain range here we would find many conglomerates and also great thicknesses of other sediments. In fact, we find neither the thick accumulation of sediments nor the mixture of sediment types that might be expected.

There has been some discussion concerning the real nature of the glacial deposits found in the Sahara. Following the publication of the 1970 discoveries, one critic wrote that we were all misled by the presence of boulders of granite and other exotic rock material in the so-called tillites. In reality, these deposits ought to be labelled 'mixtites' (mixed component rocks) or 'diamictites' (the same thing, but fancier), both terms that would honestly describe the type of formation without implying any particular mode of formation. As defined and described in Chapter 3, a 'tillite' is technically an ancient lithified 'till', and a till by definition is a rock specifically of glacial origin which contains all sorts of rock fragments, pebbles and boulders arranged just as they were deposited on the glacier bed. Dr L. J. G. Schermerhorn, a distinguished Dutch specialist with long experience overseas, submitted that a similar kind of rock mixture could be created by giant landslides (see Chapter 4). True, but surely not over thousands of kilometres? And on what was then an almost dead-flat landscape?

I have had a long useful correspondence with Dr Schermerhorn, who has published extensive and acute criticisms of the glacial theories. For this criticism we must all be grateful, for it makes us pull up our socks and take another look at some of the things we have taken for granted. Crucial in Schermerhorn's arguments (applicable, as we have seen, to the Precambrian as well as to the Ordovician glacials) is the wide range of latitude over which the so-called glacial deposits are found, whichever way you rearrange the continents by plate tectonics and polar migration. In certain areas—not in the mid-Sahara but in Morocco, for example—there are limestones containing fossils of warm-water type that were deposited cheek by jowl with the glacial rocks. Dr George Williams (of Melbourne, Australia) explains this paradox in terms of the Earth's axis changing its direction in space over long periods of time. There is nothing sacrosanct about the Earth's tilt angle; it cannot alter rapidly, but through geological time it certainly does vary. Each of the planets has a different obliquity, and the famous river valleys on Mars could well have been created at a former time when its obliquity permitted seasonal ice melting. An increased tilt of the Earth's spin axis also would exaggerate the seasonal character of the climate, giving us warm summers alternating with cold winters over a wide range of latitudes. This stimulating theory has received powerful support from palaeobotanical studies, which show that at certain times in Earth history tropical plants could live much closer to the poles, even when full allowance is made for plate tectonic movements (which for the last 200 million years are own fairly well known).

A peculiar feature of the Ordovician glacial deposits is their overwhelmingly sandy character. The characteristic tills of the UK, West Germany and North America are 'boudler clays'. Sticky, bluish-grey material, this boulder clay will change in time to a shale-like tillite, but only a few layers of shale are found in the African deposits. True, the outwash areas, where meltwater drained away from waning icesheets, created the large 'sandar' (see Chapter 3) that are well known in the British Midlands, across the North German Plain and in a belt across the USA from Nebraska to Long Island and Cape Cod. But these sandar are not made of till. The real Ordovician tillites consist mainly of boulders and pebbles interspersed in sandstone or siltstone. Why are many of these deposits so sandy?

To answer this question we must put on our palaeogeographer's cap. We have to think about the nature of the early Ordovician landscape in Africa before the ice advanced. Much of West Africa had been earlier affected by the Varangian or Eocambrian glaciation; when that ice retreated, what did it leave behind? Certainly there must have been extensive sandar. Then the Cambrian and later the Ordivician seas spread across the land, creating broad continental shelf environments where the advancing waves would churn up the loose sands to redistribute them in extensive sandy beaches and marine sand sheets. Some of these shelf sandstones are beautifully preserved today. One can clearly make out the features typical of strong wave action and currents on a continental shelf. The early French explorers found no fossils in them and concluded that these sandstones were laid down on river floodplains. Our increasing knowledge of sedimentology helps us to diagnose the bedding structures as marine, not fluvial, and a closer look has produced much useful information.

First, and most important, some thin layers of shale are found to contain beautifully preserved fossils: trilobites and brachiopods, quite clearly equivalent to those in the Caradocian Series of the British Upper Ordovician. In most places in shallow water the shells of these organisms would have been broken up by the strong wave action, but among our party of visiting specialists we were fortunate in having Professor Adolf Seilacher of the University of Tübingen, West Germany. An expert on trilobites, Seilacher is also a renowned 'tracker': he can spot trilobite footprints (collectively known as *Cruziana*) in an almost miraculous way and, furthermore, when they are well preserved, he can even tell one species from another. On successive trips he has literally tracked the trilobites all across North Africa.

Then, in many places, the surfaces of the sandstones are marked by perfectly preserved ripples and current-formed sedimentary structures. At one spot I found what I believe were traces of seaweeds. Large multicellular land plants had not yet emerged in the evolutionary scheme, but the shallow marine realm was a much more favorable ecological site for this development.

So it can be concluded that the shallow sandy formations in no way contradict the accepted environmental setting of North Africa in late Ordovician time. The icesheets must have pushed forward onto the submerged continental shelves just as they do today in West Antarctica, and not onto continental plains like those of northern Europe or North America as happened so extensively in the Pleistocene.

Ice Movement

Of great interest in North Africa is the inferred direction of ice motion. It was from the south—that is to say, from the direction of the present equator. There are several ways in which the direction of former ice motion may be established.

If you look at parts of Antarctica today you can see that the inland ice comes down to the coast in series of converging valleys. In North Africa the Ordovician palaeo-valleys converge towards the north or northwest. Just as with modern ice-filled valleys, they are commonly up to 1000m across and may be 100-300m in depth. For nearly half a century geologists in North Africa had noticed and commented on the peculiar ravines in the sandstones of the Cambrian and Ordovician that had been repeatedly scoured out and then refilled by younger deposits. In the French literature this phenomenon is referred to as '*ravinement*'. Some of the deep valleys are probably true glacial troughs cut by ice erosion. However, most of them appear to have been cut by torrential water flows.

Just how can we account for these giant flood effects? Again we go back to the Pleistocene model which has been described in Chapter 3—see page 86. The effect of the Ordovician ice loading naturally depressed the Earth's crust in the area where ice built up on the land. Peripheral to that bowl-shaped depression there must have been an elastic 'forebulge'. As the ice melted after each advance—and there were many cycles of advance and retreat—a ring of lakes would have been created at the edge of the depression, with one shore constrained by the forebulge and the other by the waning ice margin. As the ice melted back, so the crust would have risen in the depressed area, accompanied by the gradual sinking of the forebulge. Inevitably the lakes were decanted; in all probability they were catastrophically drained as the rising crust tipped them up to overwhelm the forebulge. Once each flood had started, it could cut its own channel into the forebulge to create a deep ravine. Later that ravine would become engulfed by the rising level of the ocean, and would be refilled with sediments.

In general, one can say that ice-scoured palaeo-valleys and ravines formed by the process we have just described tend to radiate away from the former ice centres. It would all be neat and tidy if the ice centres were symmetrically situated over the poles, but it is not quite so simple as that. We have already seen in earlier chapters that snow (and ice) tends to accumulate in sites that are dictated by the rules of meteorology. Favourable sites occur where moist air from the ocean meets the cold polar front; such sites tend to be located well away from the actual pole, which itself is often found to be the centre of an 'ice desert'. Ice streams generally radiate from the icesheet centres, and thus one may sometimes envisage the ice moving *towards* the actual pole position. Statistically, however, it is probably safe to say that the ice motion in general tends to be away from the poles. Thus we need to employ other types of observation in order to interpret this principle usefully.

Fieldwork at a 'micro-scale' can be helpful for the reconstruction of former ice-movement directions. This involves the small-scale inspection of ice-scoured rock surfaces. These latter often include lumpy or hummocky ground left by the scouring glacier, including the hummocks called *roches moutonnées* which we described in Chapter 3. In the Sahara, as elsewhere, these rock hummocks are frequently asymmetric, rounded and smoothed on the impact or 'stoss' side and fractured away on the lee side. The rock surface itself is usually finely striated, individual striations starting with an abrupt indentation and dying out 'downstream', as the pebble causing the scratch was gradually worn away. Occasionally there are successions of chatter marks, sequences of indented scars that likewise die away in the direction of ice motion. Successions of small fractures in the shape of crescent-moons are also useful indicators. In some of the areas examined by the 1970 expedition there were patches of fluted moraine; just like the grooved and striated bedrock surfaces these provided clear 'signals' concerning the directions of ancient ice movement.

As ice readvances over hard bedrock it tends to skim off the soil and adapt its shape in conformity with the resistant basement. What happens when it advances onto a continental shelf covered by loose sand? One could expect deep gouging out of palaeo-valleys but, on the flat interior plains of the Sahara, it seems to have moved very evenly across the sandy landscape. In such areas the ice left long parallel grooves, sometimes traceable from horizon to horizon. The grooves are commonly 50cm to 2m across with finer ridges and troughs. The underlying sandstones are not at all disturbed, and one can reason that, when the ice was advancing, the ground was probably affected by permafrost. As the ice pushed forward, the friction beneath it would create a thin film of meltwater in the manner of a skate (see page 66). There are tiny current ripples transverse or oblique to the direction of ice motion on the inner surfaces of the grooves. Similar features have not been reported elsewhere, but that may well be because few areas are known where ice has advanced over sandy deposits on flat continental shelves.

One of the French geomorphologists studying the Algerian region, Pierre Rognon, found interesting areas marked by clusters of radial fractures filled by sand. These, he concluded, were formerly cracks like those that characteristically develop in the permafrost of the high Arctic. Some of the cracks may be ancient ice-wedge casts, and a number of our party was convinced of the presence of polygonal patterns on the ground surface. Rognon also directed us to zones of basin-like scours like shallow potholes that appeared to have been scooped out of the hard frozen ground by meltwater beneath the glacier. Dr Anders Rapp, a specialist from Upsala University in Sweden, recognized them as similar to those formed during the melting of the Scandinavian icesheet of the Quaternary.

In the high Arctic of northern Canada and the Siberian tundra a number of curious features are associated with permafrost and its subsequent melting. One of the simplest features is the so-called 'tundra-crater', a small eruption or

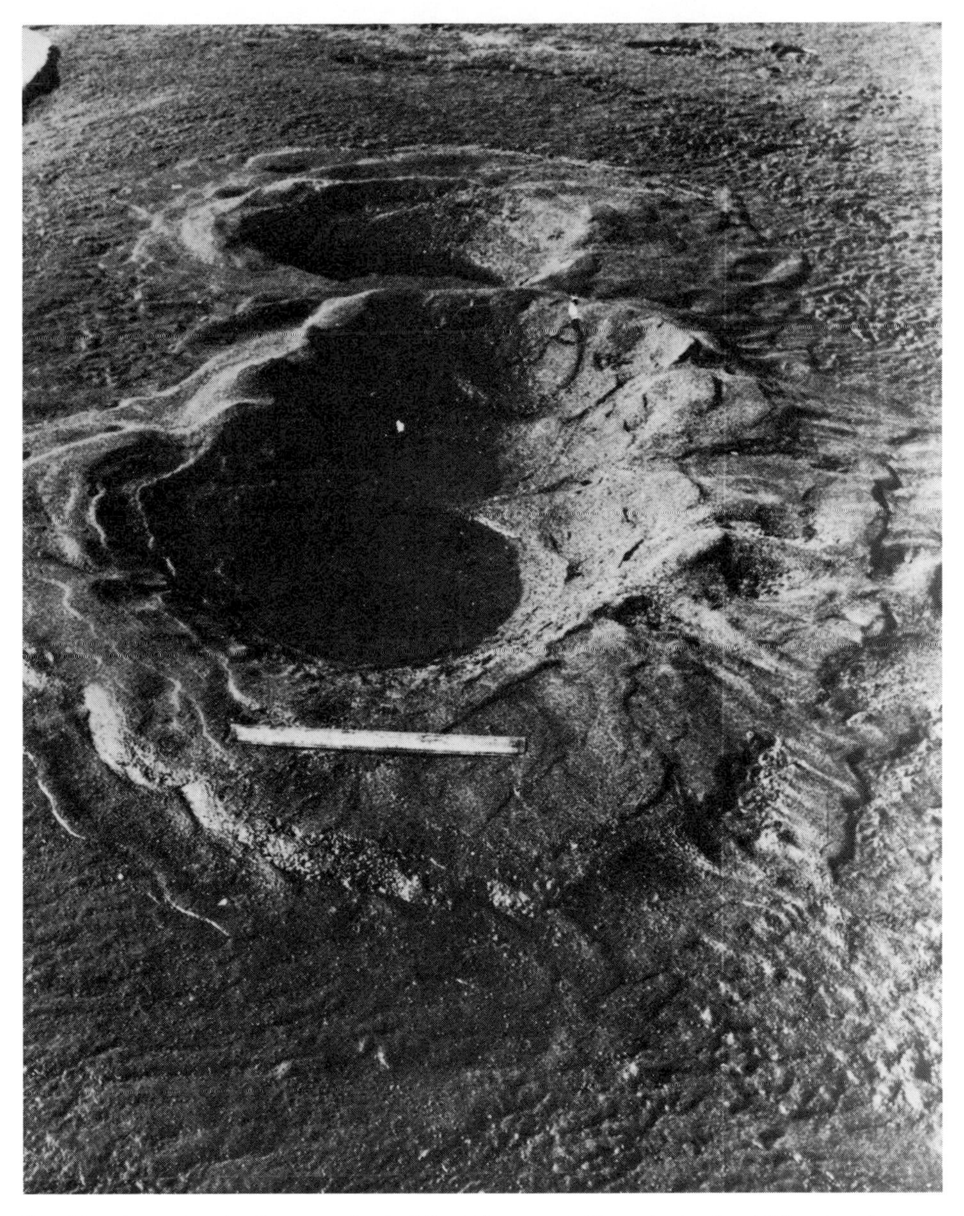

Small tundra craters in the permafrost of East Greenland (the ruler is about 30cm long). These may have formed in a similar fashion to the 'tundra craters' of the Sahara.

'boil' where melting ground ice creates a sand-laden spring like a miniature volcano. We were shown dozens of such features in the Sahara by Pierre Rognon. On average they are 1 to 5m in diameter and up to 2m high. Many of them are exquisitely preserved and may perhaps have erupted underwater and become rapidly buried; alternatively they may have been 'snap frozen' by the permafrost. Where erosion has taken off the surface layers, the feeder pipes for these 'sand volcanoes' are preserved in unmistakable fashion, for a vertical column, 20-30cm in diameter, of homogeneous silica-rich sandstone is not the sort of

Cross-bedded Upper Ordovician sandstones formed on outwash plains or 'sandar' east of the Hoggar Mountains, Algeria.

thing one sees every day in normal horizontally bedded sediments! It would be difficult to prove, but it seemed to me that large clusterings of these sedimentary eruptions might be expected in front of an advancing ice tongue, due to the pressure it would exert on water-saturated sandy formations.

As in Quaternary examples where ice readvanced over previously glaciated areas, the Ordovician icesheet usually had a smooth passage. However, here and there, a structural disturbance of the underlying formations can be seen. In some cases this takes the form of miniature folding and faulting, and in others the underlying (formerly flat-lying) stratum is broken up into large slabs which are stacked against one another, progressively stepped out, like so many roofing tiles. Some of the folding features may indicate the deformation of unfrozen sediments by over-riding ice, whereas the fault structures may be the result of the disruption of frozen materials by sub-glacial shearing (see Chapter 3). If this is the case, then the Ordovician icesheet may have been at least partly cold-based.

As the icesheets withdrew across North Africa they left large sandar just as the Quaternary ice was later to do in the northern lands. These areas are ridged by sand barriers, representing former shorelines and offshore bars, and are crossed by the channels of former outwash streams. These follow winding valleys that can be clearly traced across the flat plateaux of Africa—but, curiously,

View down into the caldera of Grimsvotn in the Vatnajökull ice cap, Iceland. This caldera in the ice is the result of ice-surface collapse connected with volcanic activity beneath the ice cap. Similar features may have occurred on the surface of the Saharan Ordovician icesheet.

the relief is now inverted. The former river gravels and sands formed favoured sites where, over the subsequent eras, groundwater has tended to accumulate a silica-rich cement. This has hardened the sands to quartzite-like resistance; under prolonged weathering and erosion the rest of the sand plain has been eroded down, leaving the former stream-bed standing out in bas-relief—the channel has become a ridge. In some places the sediments of a younger proglacial stream can be seen crossing over those of an older generation. The scene is extraordinarily vivid, and is best seen from the air, where a low-flying plane enables one to appreciate the general distribution of the local palaeogeography created by the outwash streams.

In some places there are mazes of long sinuous ridges trending towards the north. On the ground they are the size of railway embankments, and they are composed of cross-bedded sandstones. Sometimes the bedding of the sandstones is slumped on the flanks of a ridge, just as on the sides of modern eskers. Most probably, therefore, these ridges are true eskers, formed subglacially by meltwater flowing northwards towards the melting ice margin. This is apparently confirmed by the presence of fans at the northern ends of some of the ridges; most likely these were formed as the meltwater streams lost their transporting capacity on reaching the ice-tunnel exits. Again there are close analogies from glacier margins where meltwater streams emerge from ice tunnels.

One of the most spectacular glacial phenomena to be seen nowadays is in Iceland when a volcano erupts *underneath* an ice cap, such as Vatnajökull. There is tremendously rapid melting of ice and catastrophic floods occur. Somewhat similar phenomena are experienced in northern Japan and Kamchatka. As mentioned in Chapter 3, the Icelandic word for such a flood is *jökulhlaup*; in 1934 one eruption liberated fifteen cubic kilometres of water in a *jökulhlaup* lasting for only eight days.

East of the Hoggar the 1970 expedition was shown several curious ring structures up to one kilometre in diameter. At the edges the Ordovician sandstones were thrust abruptly upward and back over themselves to about 45°. In most cases the middle of the ring was covered completely with drifting sand, with no bedrock to be seen. However, inside one ring there was a low rise with outcrops of a dark fine-grained basalt, surrounded by masses of a fine breccia —a broken-up mixture of sandstone and basalt, densely fused, a typical rock of the Icelandic *jökulhlaups*. We considered the possibility that the ring structures were meteorite craters, but there were none of the pressure cones which are the typical indicators of high-velocity impacts. Some of our number thought that they might be ancient kettle-holes, or even fossil pingos; however these explanations cannot account for the presence of basalt and breccia. Dr Robert Dietz, a specialist in these 'astroproblèmes', has explored all the known Algerian examples, and he believes that the subglacial eruption hypothesis is the most plausible.

The Polar Sea

In Algeria, as one goes north from the Hoggar-Tassili Plateau area where the best outcrops are seen, the Ordovician rocks plunge gently downwards into a broad basin. The sandstones become thicker and evidently, at that time, this was a more open seaway where floating ice was still abundant. The formations at depth are quite well known because of extensive exploratory drilling for oil. The famous oil field of Hassi Messaoud has its reservoir beds in the Ordovician glacial sandstones. The porosity of the sandstones at depth is excellent, but the complications introduced by the deep channel cutting and filling (the 'ravinement') makes their commercial development rather difficult. The Ordovician glacial sandstones are, so far as we know, the only such formations that furnish a major oil field with its primary reservoir.

Continental land ice creates long parallel grooves. Even in palaeo-valleys ice tongues cannot bend round sharp corners. In places we found fine parallel grooves with an occasional 'jiggle'; in places they curved around by as much as 90° within less than a metre. These are the traces of floating ice, either pack ice or icebergs which were subject to sharp deflection under the influence of changing winds and currents. Just as a handwriting expert can stand up in court and swear to the authenticity or otherwise of a certain document, so the geologist familiar with floating ice can testify to the signature of this former Saharan ice.

Icebergs tend to drift with a deep 'root' projecting far beneath the surface. When this scrapes along the continental shelf it creates long, deep but irregular scours until it stops, digging into the sediments below as it is swung around by wind and current. Rognon showed us an area of intensely contorted sediment, seeming to betray just such a rotary scour; this seemed most likely to have been formed by an iceberg grounding.

A striated erratic block left by the melting ice in the Qasim area of northern Saudi Arabia. In this region the ice was moving from southwest to northeast.

A related phenomenon to floating ice is the creation of an 'ice foot'. This is where the sea ice freezes against the shore. But, with the rise and fall of the tide and the changing stress due to winds and currents, this beach-line segment of ice gets locked onto a tremendous load of shore material, sand, boulders, and in fact everything on the beach. Changing winds or currents can then ferry floating blocks of this ice foot along the coast or offshore, where it can melt out in a warm season. The result is an accumulation of debris rather similar to that of tillite, except that the material is often more angular, broken up by frost action, and then limited to lens-shaped inclusions in sequences of marine rocks. Distinctive examples of this rock association were also found in the central Sahara.

A Glaciated Continent

Since the time of the international expedition to Algeria, many geologists have been taking a second look at other curious boulder beds and suspicious-looking scour marks. The Ordovician glacial evidence has now been traced, for example, to Sierra Leone on the extreme west coast of tropical Africa and to Morocco in the northwest, all the way across to Ethiopia, and then across the Red Sea to Yemen, and most recently into Saudi Arabia. If this represents one continuous landmass, then it exceeded the size of the present Antarctic continent; and, if the landmasses of Ordovician time were split, the extent of the glaciation must have been even greater. Antarctica today covers nearly 14 million square kilometres; i.e., it is almost twice the size of the USA. Yet the Ordovician glacier ice once extended over at least 20 million square kilometres.

By analogy with the northern-hemisphere Quaternary glaciations, there must have been several different icesheets over this vast and for the most part nearly flat landscape. During the main glacial stages these icesheets probably coalesced. Extensive areas of 'ice deserts' must have existed, for there is no way that the flow lines of moist warm air could have passed far into the interior of this great landmass which constituted a large fraction of Gondwanaland. Some of these ice deserts may have been vast expanses of thin snow, supplied with inadequate moisture for them to develop into icesheets. Elsewhere in the interior there must have been large areas of permafrost in places separating the main icesheets.

In some of the marginal areas, the glacial formations are apparently of marine origin. That is to say, the continental icesheet edges could not have been far away, but the formations now preserved were deposited on the submerged continental shelf. Good examples are the glacial sediments of Sierra Leone, studied by M. E. Tucker and P. C. Reid of the University of Sierra Leone. These beds correlate with similar ones in Senegal and Guinea and consist primarily of about 150m depth of laminated mudstones; they comprise seasonally banded layers of coarse and fine sediments that suggest frozen seas for half the year. Like the Precambrian glacial deposits they contain dropstones transported by floating ice and released into relatively deep water. There are ravines and channel fillings. Beneath the glacial beds there are cross-bedded sandstones just like those in Algeria, 3,000km away.

Probably the first positive report of Ordovician glacial deposits was in Mauritania, by two French geologists, J. Sougy and J. P. Lecorche, in 1963. Thanks mainly to the work of J. Destombes, the ice age features have been traced extensively through the Anti-Atlas of Morocco. G. Bronner and J. Sougy discovered them again farther south in the former Spanish Sahara. Then, on the other side of the Sahara, glacial traces were identified in the Ennedi on the borders of Chad and the Sudan by S. Beuf and his colleagues. A probably late Ordovician age was then attributed to glacial rocks in Ethiopia by D. B. Dow, M. Beyth and T. Hailu (in 1971). Next, Ordovician glacial traces were picked up again across the Red Sea in Yemen. And now, most recently, beautifully preserved Upper Ordovician tillites have been reported by H. A. McClure in Saudi Arabia (1978), well over 6,000km from Sierra Leone; there are also Caradocian trilobites, *Cruziana*, *Skolithos*, deep channel fills and comparable features there.

The Ordovician rocks are generally absent in central Africa but they reappear in South Africa and Namibia. In spectacular outcrops on the sides of Table Mountain behind Cape Town there is a 750m-thick sequence of littoral and lagoonal sandstone and shaley deposits that rest on a peneplained Precambrian basement, and are overlain in turn by a tillite called the Pakhuis Formation. In 1967 I. C. Rust reported finding a glacially striated pavement here, as well as the seasonally banded layers (varves) of a glacial lake deposit. There are fossil seaweeds and marine brachiopods. These formations have long been known but had been incorrectly identified as from the Devonian. The fossils are not as perfect as those of North Africa, but suggest an Ashgillian (end-Ordovician) age rather than Caradocian. Inasmuch as the magnetic pole was shifting to the south in this period, it would not be quite proper to claim that the Pakhuis Tillite belonged to the same icesheet as the Saharan tillites, but, as pointed out above, in an ice age as prolonged and as extensive as this, one would expect to

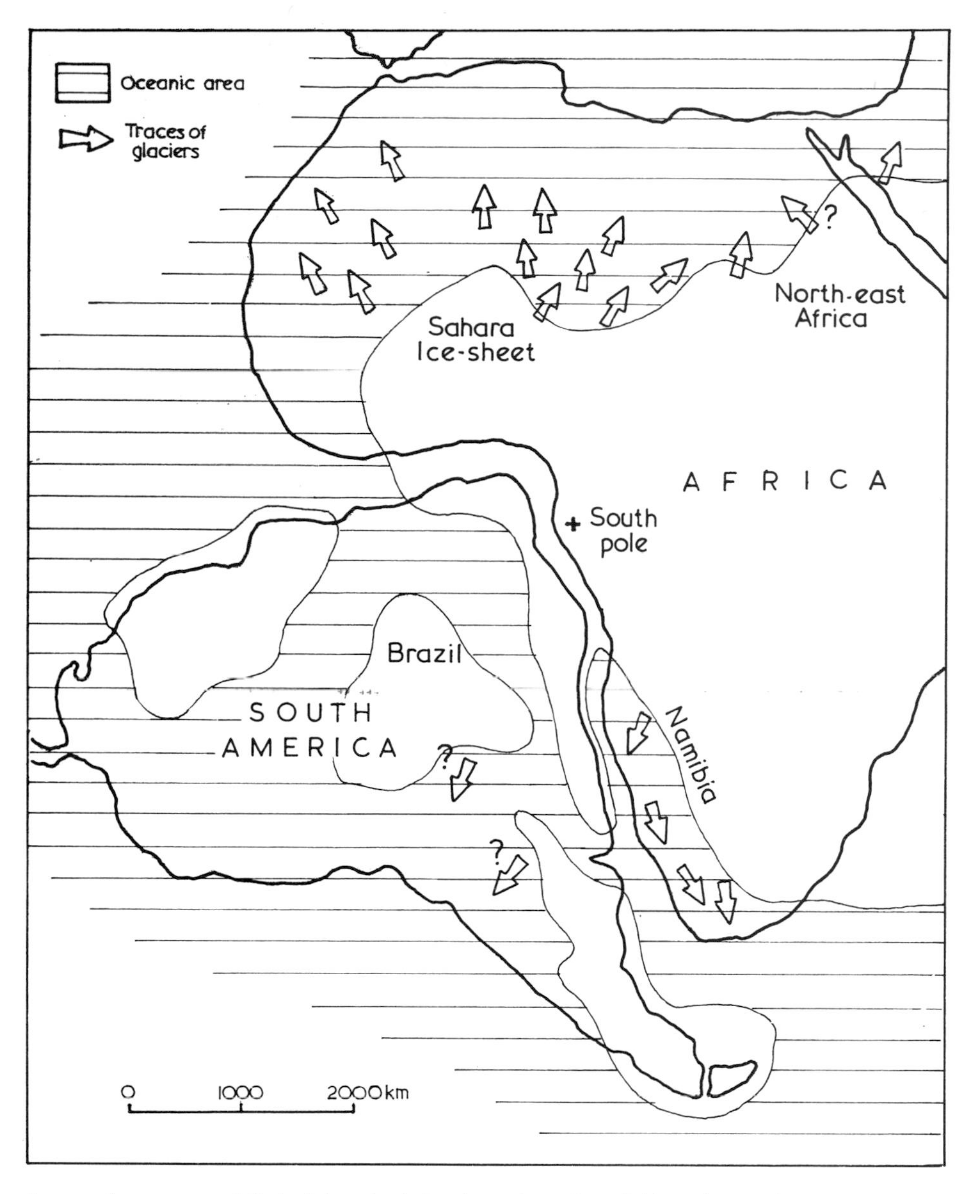

Areas where traces of Ordovician glaciation have been found. Note that the land areas were rather different from those of today. The ice-covered area must have been considerably larger than that of present-day Antarctica. Probably there were multiple ice centres, as in the northern hemisphere during the last glaciation. There may have been independent ice centres in West Africa, Namibia, Brazil and northeast Africa.

(*Overleaf, left*) The icesheet margin in East Africa may have looked something like this. In this vertical air photo of part of the edge of the Greenland icesheet we can see bare rock areas, patches of thick till, meltwater lakes and extensive areas of fluvioglacial outwash deposits beyond the ice edge.

(*Overleaf, right*) The icesheet margins in northwest Africa and Namibia may, at certain times during the ice age, have looked like this. This vertical air photo from Adelaide Island, Antarctica, shows a grounded ice margin from which many large icebergs are being calved. Most of the deposition is of glaciomarine materials dispersed by floating ice.

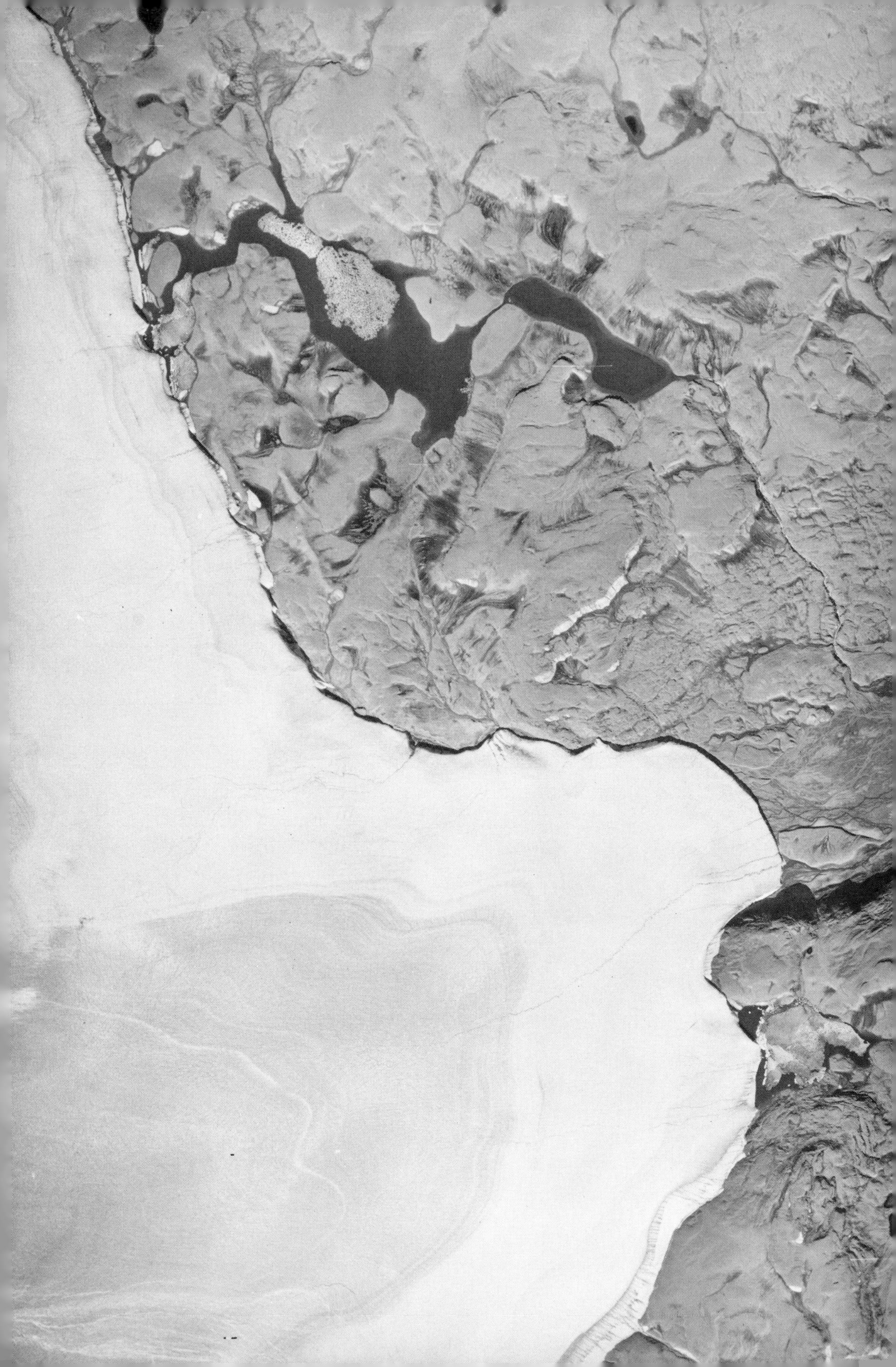

find traces of a number of different icesheets and a number of different glacial stages. The distance from Cape Town to Morocco is 6,000km.

Beyond the limits of Africa there are numbers of Upper Ordovician boulder beds (properly called 'diamictites') that are less well preserved, but where the evidence is suggestive of glaciation. Each of the places appears once to have been attached to North Africa prior to plate-tectonic rifting and sea-floor spreading: Nova Scotia, Boston (USA), Spain, Sardinia and France. Each of these sites still needs work to be done in order to make absolute confirmation.

South America was also part and parcel of Africa in Ordovician times, and one naturally looks to see if there are matching formations. On a recent trip to Brazil, I was able to do some checking, with the kind help of Professor A. C. Rocha-Campos of the University of São Paulo. In Brazil there is the Trombetas Formation in the Amazonas basin which is certainly glacial in origin, but the age is only loosely established as Ordovician. Then, in Bolivia, in the Altiplano and Subandina regions, there is the Cancaniri Formation, but similarly the age has been only tentatively determined. At present, therefore, the conservative thing is to say that South America is a 'probable', but that the evidence obviously needs closer investigation.

A note of caution should be introduced concerning a commonly reported Silurian glaciation from Africa and South America, and sometimes even from Western Europe. These reports come in particular from French and South American geologists who up until quite recently did not recognize the Ordovician as a separate period; to them it was the 'Silurien Inferieur' (lower Silurian) or 'Eosiluriano'. Quite possibly, therefore, the 'Silurian glaciation' of the literature is simply what one might call a 'terminological Ice Age'—that is, it is merely the Ordovician ice age under another guise.

The Structure of the Ice Age

Although we have only a decade or so of observations to work from, attempts have already been made to clarify the main events of the Ordovician ice age. I pointed out at the beginning of this chapter that the area affected by ice was drifting slowly across the south pole, so that in effect the pole 'migrated' from North Africa towards the Congo Basin and thence towards southwest Africa. During the late Ordovician the south pole was located not far from the south-western part of the Sahara, and this was the time (around 450 million years ago) when the ice age seems to have been at its peak.

From what we know about the later Ordovician elsewhere in the world, there were a number of rapid marine transgressions across the land and regressions from it again; these may have been related to the interglacial and glacial stages of the ice age. The Saharan evidence suggests that there were at least three (and probably many more) advances and retreats of the ice, but it is not known whether these were glacial-interglacial cycles or simply related to stadials and interstadials of the northern icesheet margin. The sequence in a typical cycle is as follows: grooved and glacially eroded surface (at the base); basal tillite; glaciomarine deposits including dropstones; and finally (at the top) waterlain siltstones, sandstones and conglomerates which may have had their origin as material carried from the glacier by outwash streams. In places the sandstones contain units with gigantic 'rippledrift' bedding which is characteristic of

Multiple tillite horizons of Upper Ordovician age exposed in a cliff to the east of the Hoggar Mountains, Algeria. These probably indicate repeated expansions and contractions of the icesheet across this area.

conditions of very high discharge of water—in terms either of velocity or of the amount of material carried along by the water. And in places the sandstone and conglomerate surfaces appear to have been affected by permafrost, as we saw on page 140. A reasonable climatic interpretation of one of these cycles would be as follows:

TOP

6. *Permafrost features:* Surface affected by permafrost. Cooling leading to next glaciation?
5. *Emergence of land surface:* Further retreat of icesheet. Interglacial?
4. *Water-lain sediments:* Shallowing sea (possibly due to the land rising as the glacier melts and the pressure is consequently reduced). Large sandar beyond ice margin.
3. *Glacio-marine deposits:* Disintegration of icesheet. Inundation by sea. Possibly ice shelves and pack ice present.

2. *Basal tillite:* Deposition by icesheet (shortly after maximum of glaciation?).
1. *Glacially eroded surface:* Ice advance, sometimes by a wet-based icesheet. Removal of interglacial or interstadial deposits?

BASE (Usually marine sandstones marked by ripples, crossbeds and tracks of organisms, and with evidence of strong tidal currents.)

This sequence of deposits and events is of course quite similar to that mentioned on page 120 for the Port Askaig tillite sequence in Scotland; it appears that in Ordovician North Africa, just as in Precambrian Scotland, the icesheet was quite unstable, with marine inundations accompanying the rapid if not catastrophic wastage of the icesheet margin. We can imagine that the environment must have been very similar to that of present-day West Antarctica, where the icesheet is grounded beneath sea-level and where the catastrophic melting of both the icesheet and its fringing ice shelves is a distinct possibility (see page 201). As John Andrews describes later in this book, the disintegration of the Laurentide and Scandinavian icesheets during the present ice age seems to follow exactly the same pattern.

From the above points we can make the following speculations about the structure of the Ordovician ice age and about the glaciological characteristics of its icesheets:

(*a*) there were probably at least three, and possibly as many as twenty, glacial stages separated by interglacials;
(*b*) the northwestern margin of the Saharan icesheet was probably for the most part warm-based; and
(*c*) the northwestern margin was an unstable 'maritime' margin characterized by the catastrophic disintegration of icesheets and ice shelves.

End of the Ice Age

To students of ice ages in general, the end of a glacial interval is a question of profound interest. In the central Sahara the structural position is extraordinarily stable, almost flat, and there is no possibility of any tectonic disturbance. The broad expanses of tillites, grooves and outwash sands are neatly covered over by Silurian shales resulting from the sea transgressing across the land. These Silurian beds, similar to those of Central Wales, are marine and rich in fossils, and they rest directly on glacial formations with no evidence of any long interruption. The fact that these beds are shales, without any sandy layers, proves a remarkable point. The Silurian sea must have 'drowned' most of North Africa in one tremendous inundation. Shales come from clays, and clays are transported into open ocean regions in suspension: if the coast was advancing landwards only gradually, erosion would have stirred up the soft glacial sands and incorporated them into new coastal and shallow-water formations.

Such a large-scale 'drowning' probably involved a sudden rise of sea-level of at least 300m, possibly as much as 500m. It is hard to explain such a rise on the basis of glacial meltwater alone. (If we melted all of the Antarctic ice today, the sea-level rise would be only 65m.) Thus some catastrophic event is called for. While it is not yet tested, I am attracted by the idea that the ice's eccentric loading of the Earth's crust and its subsequent redistribution as water caused a sudden change in the polar axis, which would lead to readjustment in the actual shape of the globe. A sudden shift of palaeolatitude towards the equatorial bulge would cause an instantaneous rise of sea-level. It has been calculated that a polar shift of only 1° would raise sea-level at the equatorial bulge by 375m, which would be just about right for explaining the great post-glacial inundation.

Following the Ordovician ice age there was, according to the known evidence, no further glacial action anywhere in North Africa, apart from the Quaternary mountain glaciers of the Atlas. According to the palaeomagnetic record, the 'Gondwana' pole now migrated rapidly to the south and, by Permian time, reached Antarctica, which was then still attached to South Africa.

Some traces of suspected Silurian and Devonian glacial formations have been reported from different parts of eastern South America and from the western coast of southern Africa, but these have not been confirmed and do not seem to have been associated with any major glaciations. Possibly they were formed during a period of mountain glaciation which continued in favourably located upland areas for several millions of years after the great lowland icesheets had disappeared.

Following the demise of the Ordovician icesheets, there was a long mild interval in the Earth's climatic history. It was to end about one hundred and fifty million years later, with the advent of a new 'winter of the world'; the glaciers were to advance again with the inception of the Permo-Carboniferous ice age.

RHODES FAIRBRIDGE

6

The Great Ice Age: Permo-Carboniferous

The late Palaeozoic era was a time of profound change in the character of the global environment. The surface of the land was colonized on a large scale by both plants and animals and there was also a prolonged and intensive ice age which made its greatest impact upon the supercontinent of Gondwanaland. During the Carboniferous there was a gradual cooling of world climate which culminated, about 300 million years ago, in extensive icesheet glaciation, large areas of permafrost in both hemispheres, and widespread ice shelves and areas of floating sea ice. At the same time there was a gradual fall of sea-level. The irregular withdrawal of the sea from the continental landmasses was caused by tectonic activity and crustal uplift in some areas, but it may also have been linked to the build-up of the southern hemisphere icesheets.

The impact of the Permo-Carboniferous ice age was far greater than many geologists and palaeontologists have recognized, and the process of glaciation had dramatic effects even in equatorial regions.

A simple illustration of this comes from a study of the sediments which were accumulating on the edges of the ocean basins. In middle and late Carboniferous times there were many rapid marine transgressions and regressions which affected the course of sedimentation in the world's shallow seas. These rather violent movements of sea-level were quite possibly of glacio-eustatic origin (see Chapter 3), coinciding with episodes of icesheet growth and disintegration during the stages of the ice age. In many sequences of sedimentary rocks which contain coal beds (for example, in both North America and Western Europe) there are 'cyclothems' or repeated sequences of rocks such as limestone, shale, sandstone and coal.

It has been suggested, quite reasonably, that these cyclothems were linked with sea-level changes resulting from icesheet surges. Each one consists of a sequence of up to nine different rock types. In the USA the lower beds are often of non-marine origin, culminating in beds of coal laid down close to sea-level. Above the coal beds there are limestones and shales which seem to indicate rapid submergence, and these are followed by rocks laid down in or near river deltas with the building up of the sea floor towards the new sea-level. The peats and terrestrial swamp forests which were later compressed and turned into coal beds mark the approximate positions of sea-level immediately before each submergence. The top of each coal bed and its contact with overlaying shale or limestone marks the 'culmination' of each cyclothem.

In the Pennsylvanian succession of the USA (representing some 40 million

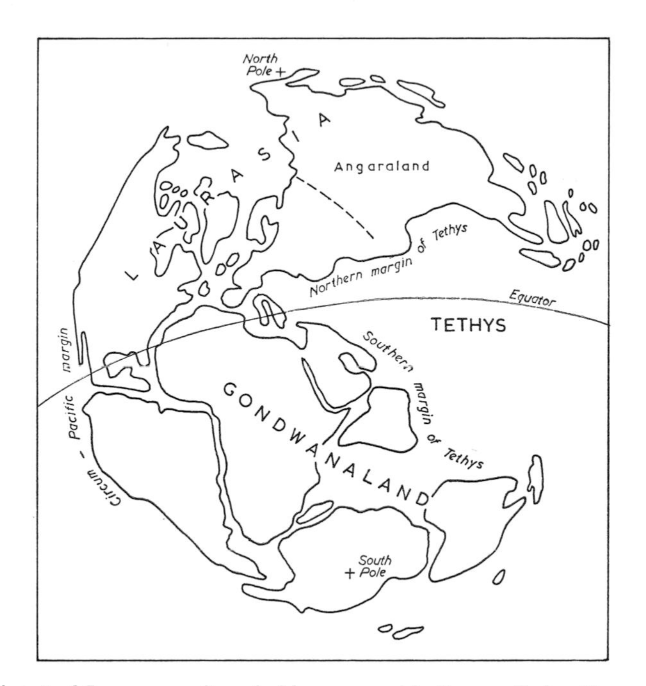

The supercontinent of Pangaea as it probably appeared in Permo-Carboniferous times. In the west the two old supercontinents of Laurasia and Gondwanaland were fused together near the equator; in the east they were separated by the Tethys.

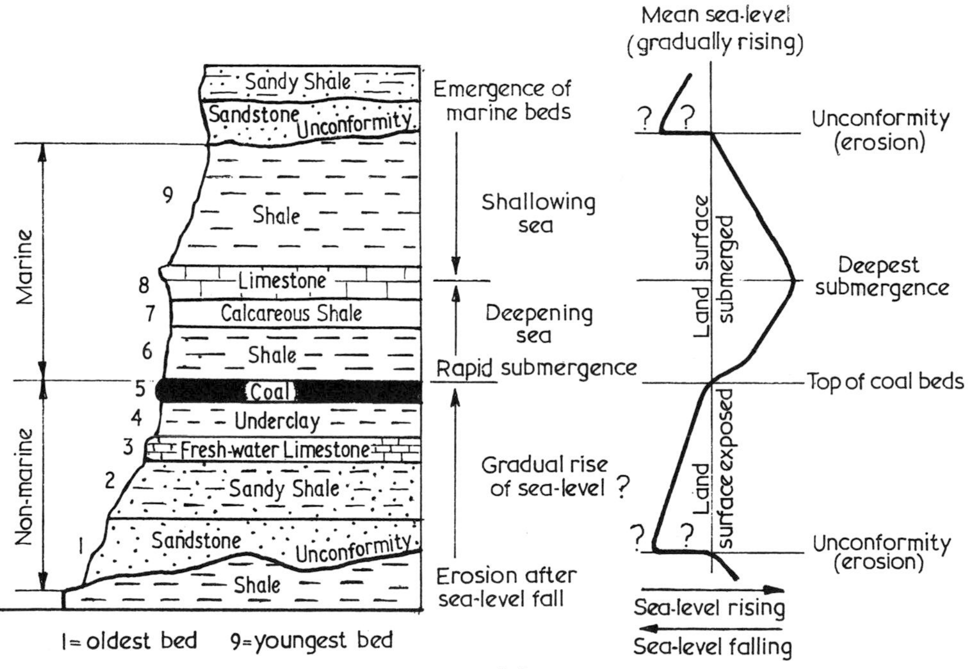

A typical cyclothem from the Carboniferous Coal Measures of North America. The graph on the right shows how the various beds can be related to rising and falling sea-levels.

years of depitioson) there are about 100 cycles of sedimentation. Knowing what we do about the violent oscillations of sea-level during the present ice age, it is quite logical to assume that most of these cyclothems may have been linked in a general way with the waxing and waning of the Gondwanaland icesheets and more particularly with repeated icesheet surges.

The Permo-Carboniferous World

Before we examine the evidence for the Permo-Carboniferous ice age it is worth looking at the global context in which this prolonged glaciation occurred. As pointed out in Chapter 1, the Carboniferous period (which lasted from about 345 million until 280 million years ago) was a time of continental 'clustering'. At the beginning of the period there were three distinct landmasses which spread across most of the climatic zones of the Carboniferous world. These climatic zones were probably not all that different from those of the present day, although overall world-wide temperatures may well have been slightly lower. Also, in contrast to the present day, the arrangement of climatic zones would have been affected by the large amount of land in the middle and high latitudes of the southern hemisphere and the limited amount of land north of the Equator. Temperature differences between the polar regions and the tropics may well have been more extreme than they are today.

In the northern hemisphere the continent of Angaraland occupied mainly boreal latitudes, although part of it lay above 75°N. Originally a gulf separated Angaraland from Laurasia, a considerably larger landmass which stretched across a wide area from about 15°S to about 40°N. This landmass, which included a number of offshore islands, incorporated most of modern North America (excluding the Cordillera, which had not yet been created), Greenland and Europe west of the Urals. Gondwanaland, the largest of the three ancient supercontinents, stretched from near the Equator into the south polar regions. Virtually the whole of Antarctica, in addition to parts of South America, South Africa, India and Australia, lay above 55°S in latitudes which were favourable for the onset of glaciation. During later Carboniferous time Angaraland drifted towards Laurasia, closing the old sea gulf. Gradually, as Laurasia rotated in a clockwise manner, the two supercontinents fused together along the line of the Ural Mountains. By the mid-Permian period one vast supercontinent stretched from pole to pole. This is generally referred to as Pangaea, a crescent-shaped landmass which survived for almost 100 million years until it broke up during the Jurassic period. The rest of the surface of the planet was covered by a single great ocean, part of which, Tethys, extended into the heart of the supercontinent.

The Permo-Carboniferous world was vastly different from that of the present day, in spite of the fact that clearly defined climatic zones were already in existence. There was a rich plant life in the oceans, but the land surface was largely barren, at least during early Carboniferous time. As the landmasses moved together with the 'fusing' of Pangaea, old geosynclines were uplifted to form mountains and many marine environments were eliminated. Ocean currents were diverted, again leading to drastic environmental changes in parts of the world ocean. And in many areas well away from the new mountain ranges gradual land uplift led to the exposure of wide shelves or coastal platforms around the continental margins.

In favourable localities the colonization of the land by plants, which had accelerated during Devonian times, was continued. There were large forests of tree ferns, scale trees, scouring rushes and seed ferns. There were some primitive pinelike conifers, and a few species of scale trees and seed ferns grew to heights exceeding 30m. Gradually the old species of seedless trees were replaced by the more advanced seed-bearing ferns and gymnosperms. Before the end of the Carboniferous period true conifers were thriving.

The vast swamps and coastal forests of the tropics led, in some areas, to the accumulation of thick peats, which have since been converted to coal beds. But in general it seems that the hot wet tropics were not all that suitable for coal formation; probably the decomposition of vegetation was too rapid for great thicknesses of peat to accumulate. In contrast, the late Carboniferous and Permian coal beds seem to have accumulated in the middle and high latitudes of both hemispheres. Some of the Permian coals are actually associated with tillites, and they contain tree trunks with prominent growth rings which imply slow growth in an environment similar to that of the muskeg or northern boreal forest zone of Canada or Siberia of the present day.

In many areas the land vegetation was decidedly less lush. Over much of Gondwanaland glacial and periglacial conditions prevailed for many millions of years, and beyond the ice edge there must have been polar deserts and wide expanses of tundra vegetation. The *Glossopteris* flora of Gondwanaland has been closely studied, and it seems to have been dominated by plant species which were adapted to cold climate conditions. This flora was widespread over virtually the whole of Gondwanaland, and traces of it have been found in Carboniferous and especially Permian rocks from large numbers of sites within and even well beyond the area affected by the Permo-Carboniferous icesheets. In the equatorial belt there were some areas where desert conditions persisted until Permian times, with dune sandstones, coarse conglomerates and evaporite beds indicating land

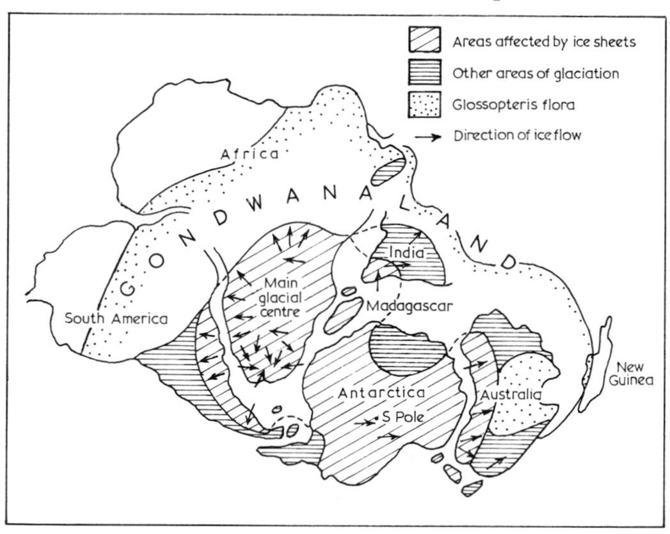

The Permo-Carboniferous glaciation of Gondwanaland. Not all areas were affected by glaciers at the same time; in general, South America and southern Africa were affected earliest, and eastern Antarctica and Australia latest. Within the area of *Glossopteris* flora many regions were intermittently submerged by the sea.

surfaces which were largely devoid of plant life. The land surface also seems to have been barren in some mid-latitude areas; runoff of rainfall was rapid, especially where new mountain chains and newly uplifted platform areas were subjected to attack by the forces of denudation. There were myriads of streams and rivers involved in the transport of rock debris. Thick conglomerates, grits and sandstones were laid down on the land surface in basins and valleys and along the continental coasts.

The oceans of about 300 million years ago provided a range of conditions which were probably broadly similar to those of the present day. There were muddy seas fringed by deltas and swamps of terrestrial vegetation. There were seas affected by pack ice where glacio-marine sediments accumulated gradually on the sea-floor. And in tropical areas there were widespread areas of clear-water marine conditions, where the warm waters were ideal for the formation of coral and algal reefs and shell banks. Sometimes, in shallow bays within the tropics, evaporation rates were so high that evaporites were crystallized on the sea bottom. In the deeper ocean regions marine trenches accumulated thick sequences of sands and muds containing plant and animal remains; these dominantly gray and black sandstones and shales were gradually distorted and folded as the old geosynclines were filled and then uplifted. Seaweeds were widespread in the shallow ocean areas, and in the warmer shelf seas there was a rapid evolution of creatures with calcareous shells and skeletons. Corals, brachiopods, molluscs and crinoids abounded, and their fossil remains are abundant in limestones of Carboniferous and Permian age.

The Permo-Carboniferous environment was, as indicated in the foregoing paragraphs, greatly affected by the processes of plate tectonics and mountain-building. As in all global episodes of continental 'fusion', the collisions of plate margins resulted in a number of orogenies which created new mountain ranges. In middle and late Carboniferous times European Russia was affected by the so-called Hercynian or Armorican orogeny, and this episode of mountain-building also affected other parts of Europe at the end of the Carboniferous period. The collision of the Ancestral Siberian and Ancestral European plates caused the uplift of the Ural Mountains, and the collision of Laurasia with Gondwanaland caused the uplift of the Hercynian mountains of Europe. In the British Isles, France and Spain the Hercynian structures in the Palaeozoic rocks are aligned broadly east-west, indicating a 'squeezing' from north and south. In North America a new mountain range was formed along the contact between Laurasia and the northwestern edge of Gondwanaland. An old geosyncline stretching from Newfoundland and Nova Scotia southwestwards to the Appalachians and thence towards Texas and New Mexico was filled with sediment and then intensely folded and uplifted. There were several 'pulses' of mountain-building along this plate contact, the main phase of uplift being that generally referred to as the Appalachian or Alleghenian Orogeny. The deformation of rocks and the creation of high mountains was probably at its peak during early and middle Permian times. There was also folding and deformation, with some mountain-building, along the oceanic edges of Pangaea. For example, there were many pulses of uplift along the western margin of South America during the Late Palaeozoic, and there was uplift along the northern margin of Tethys during the middle part of the Permian period.

The creation of large mountain chains during Permo-Carboniferous times had a number of important consequences. In the first place, there were widespread volcanic eruptions and outpourings of lava along many plate margins. Secondly, with the broad cooling of climate which followed the end of the Devonian period, highland snowfields and ice caps were able to form in mid-latitude and even equatorial areas. Patterns of atmospheric circulation must also have been affected. Rates of surface erosion were speeded up within and on the flanks of the new mountain chains. Quite possibly these environmental changes helped to reinforce the climatic cooling of the Permo-Carboniferous ice age. The 'clustering' of the supercontinents, the great continental area in the south polar regions, and the altered pattern of ocean currents must also have contributed in large measure to the global cooling. While there may still be doubts about the precise 'trigger' mechanisms which led to the onset of the Permo-Carboniferous ice age, there can be no doubt about either its persistence or its intensity. The geological record contains abundant testimony to a prolonged and yet dramatic event which we can with confidence refer to as 'the Great Ice Age'.

Evidence for the Ice Age

Some of the best evidence for the Permo-Carboniferous ice age comes from the southern part of the South American continent. Quite possibly there were alpine glaciers in the uplands of Bolivia and Argentina during Devonian times, but as the climate cooled around 345 million years ago glaciers became more extensive in many areas. In Bolivia there are three distinct glacial horizons. The first and possibly the second of these are of early Carboniferous age; tillites of a similar age are known also from Argentina. The Bolivian glacial deposits appear to have been laid down on land, but those of Patagonia and western Argentina are difficult to interpret. Lawrence Frakes and John Crowell, who have studied the Permo-Carboniferous glacial deposits more intensively than other geologists, consider that they are partly glaciomarine in origin and partly redeposited by streams and marine processes. In any event the rocks are clearly indicative of widespread glaciation over a north-south distance of at least 2,600km. Probably most of the glaciers originated in the highlands and descended eastwards onto the mountain forelands as piedmont lobes and westwards as floating glacier tongues and ice shelves in broad embayments and fjords.

The glacial features of the Parana Basin in Brazil are quite different. They cover an area of at least 1.5 million square kilometres, and in places the glacia beds are 1,600m thick. Many of the classic features of glaciation are preserved here, including tillites, striated rock platforms, glacially eroded valleys, glacio-marine sediments with dropstones, faceted pebbles, *roches moutonnées* and eskers. Towards the west, where the glacial deposits outcrop along the edge of the Andes, there are signs that the glacial materials have been reworked, but elsewhere the glacial sequences are *in situ* and beautifully preserved. The so-called Itararé beds of the Parana Basin and southern Brazil are mainly land-derived, and in places there are six separate tillites. Some parts of the Parana Basin were intermittently submerged beneath sea-level in Permo-Carboniferous times, for there are a number of layers of glacio-marine sediments intercalated with the true tillites.

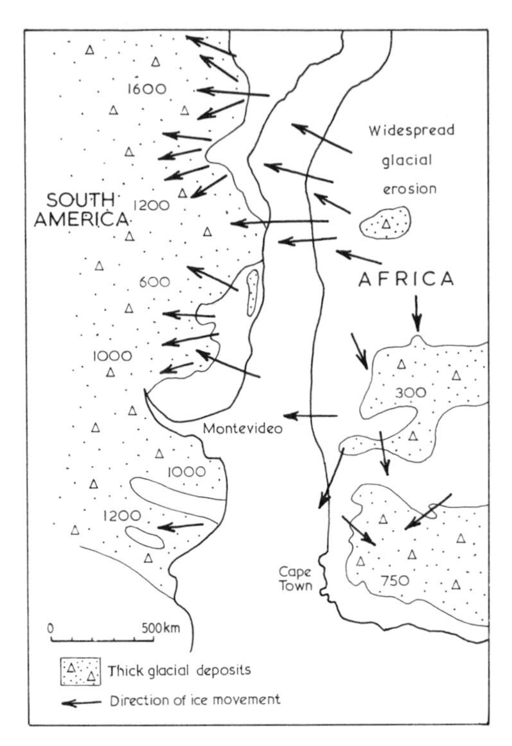

Glacial deposits and directions of ice movement (inferred from striae and other features) in southwest Africa and eastern South America. In Permo-Carboniferous times the continents were joined together. Numbers refer to the thicknesses of glacial deposits in metres.

The glacial deposits of southern Brazil seem to span quite a long period of time, from about 320 million years until about 270 million years ago. During this period of 50 million years there appear to have been at least 17 glacial/interglacial cycles. However, the greatest extent of the icecover in the Parana Basin occurred about 280 million years ago, and it has been suggested that here the ice age lasted for no more than 5 million years.

There is no doubt that the icesheet which was responsible for the glaciation of eastern South America had its source area somewhere off the present coastline. The direction of ice movement can be inferred from striated rock pavements, directions in which erratic boulders have been transported, and folds and thrusts in the sediments beneath the glacial deposits. Always the evidence shows that the ice moved broadly from east to west.

Before the acceptance of Wegener's continental drift hypothesis this would have caused a problem of interpretation, but when we look at the map of Permo-Carboniferous Gondwanaland it is entirely reasonable to suppose that the ice which affected Brazil and Uruguay came from a source area in southern Africa. This hypothesis is supported by many different types of field evidence. For example, the tillites of Brazil contain erratic blocks of distinctive coloured quartzite, dolomite and chert which cannot be matched with any known source rocks in Brazil. But these erratics *can* be matched in the tillites of southwest Africa, and in some cases the source rock has also been located there. The glacial deposits of southwest Africa are rather sparse, and they are preserved for the most part in ancient valleys. On the other hand, there is abundant evidence for

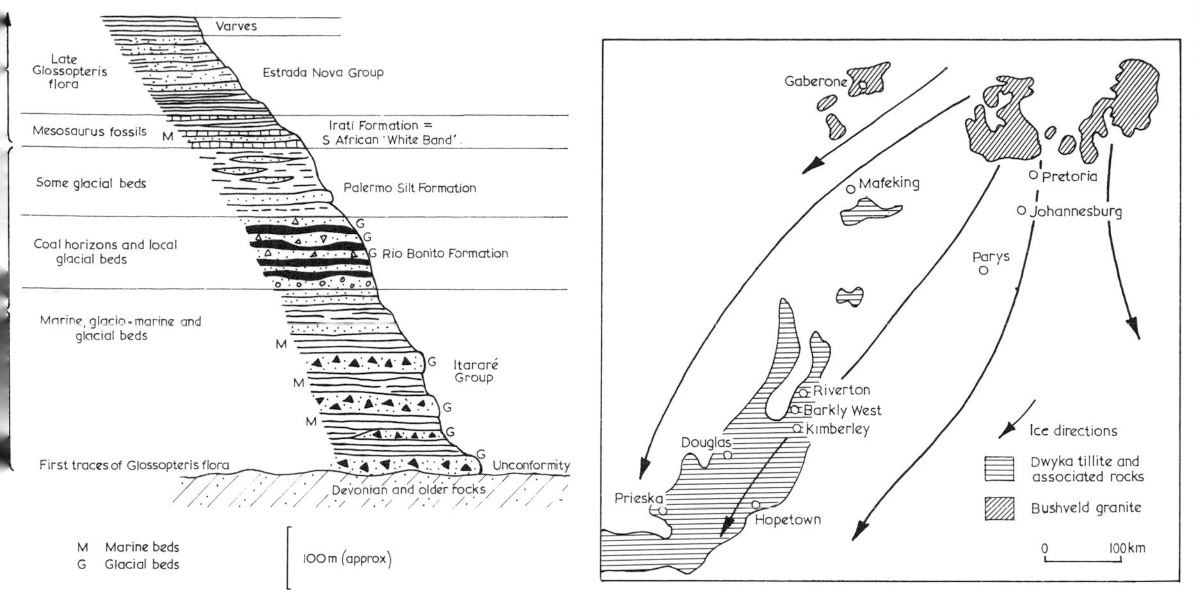

(*Left*) Generalized diagram showing the Carboniferous and Lower Permian formations of part of southern Brazil. Most of the glacial deposits are found in the Itararé Group of rocks.

(*Right*) Directions of ice movement across part of southern Africa, inferred from the transport of Bushveld granite. Some of the erratics are contained in the Dwyka tillite in the south-western part of the map area.

widespread ice erosion in the form of *roches moutonnées* striated rock pavements and polished and striated valley sides. The direction of ice movement indicated by these features is again broadly east to west.

By far the most reasonable interpretation of the evidence from both sides of the South Atlantic is that ancient southwest Africa was affected by an actively eroding (in other words, warm-based) icesheet which dumped its load further to the west in what is now South America. The interbedding of tillites and glacio-marine sediments strongly suggests that the western part of the icesheet was unstable and maritime in character, just as the West Antarctic icesheet is today (see Chapter 7).

The Dwyka tillite of southern Africa and Madagascar was first described in 1909, and since then both its Permo-Carboniferous age and its glacial origin have been widely accepted. The old tillite and other glacial deposits lie at or near the base of the Karroo system, in widely separated localities extending for over 1,600km east to west and 1,200km north to south. The tillite thickens southwards, reaching a thickness of well over 1,000m in the southern Karroo. Associated with it are striated pavements, glacial valleys infilled with glacial deposits, fluvioglacial horizons, and glacio-marine and marine horizons which contain marine fossils of early Permian age. The marine beds are found for the most part in southwest Africa and the Cape Province, but in Natal and Transvaal the beds are clearly terrestrial; the Transvaal was probably one of the main highland areas of ice dispersal.

In the Cape Province the Dwyka Series has been subdivided as follows:

Upper Shales	220m	Lower Permian
Tillite	830m	Upper Carboniferous
Lower Shales	250m	Middle-Upper Carboniferous

The Lower Shales are pale green in colour and contain some fossils. The Tillite is for the most part without fossils, but there are some horizons with traces of *Gangamopteris*, one of the typical cold-climate plants of Upper Carboniferous times. The Upper Shales are typically grey-green in colour, and most horizons contain traces of a *Glossopteris* flora. In the shales there is one particularly prominent marker horizon, a carbonaceous 'White Band' which contains fossils of the aquatic reptile *Mesosaurus*. This sequence of Upper Shales is of Permian age, and it dates the end of the Dwyka Glaciation.

Outside the Cape Province it is relatively uncommon for the Dwyka Tillite to be contained between marine horizons. In Natal, for example, the glacial deposits rest directly upon glaciated pavements which are cut across the bedding of older rocks. These pavements have superb striations, chatter-marks and crescentic gouges, and appear to have been deeply eroded by overriding ice. In places there are *roche moutonnées*, and on the Kaap Plateau of South West Africa there are magnificent glacial troughs which contain abundant evidence of deep erosion by ice. The tillites, wherever they occur, contain faceted and striated boulders and stones, many of which are oriented parallel with the direction of ice movement. Detailed studies of till 'fabric' have also revealed that the smaller particles in the ancient tills are aligned in the ice-flow direction. Most of the stones in the tillites are from local sources; however, there are some far-travelled erratics and in some instances stones have clearly been derived from older glacial deposits overridden by readvancing glaciers.

In general, there are three different types of glacial deposit: massive unbedded tillite of terrestrial origin; bedded tillites with dropstones and marine fossils; and varved beds with dropstones which apparently originated in lakes. In recent years it has been realized that there are also quite widespread fluvioglacial horizons, or horizons resulting from the 'reworking' of old glacial deposits by interglacial and post-glacial rivers.

The events of the Dwyka glaciation have still not been worked out in any great detail, and the evidence from different parts of southern Africa often appears contradictory. In South West Africa there is evidence for two distinct phases of glaciation with a time interval between them, but elsewhere there are signs of at least four major advance/retreat cycles. Since the Permo-Carboniferous ice age in this area seems to have lasted for at least 20 million years, and since the position of the south pole was migrating throughout this period, complications in determining both the timing of glacial and interglacial stages and the directions of ice movement are only to be expected. The South African geologist Alexander Du Toit, who studied the Dwyka glacial deposits for many years, suggested that there were four major centres of ice disposal: (*oldest*) Namaland (north-south flow); Griqualand (radial flow); Transvaal (probably radial flow); Natal (northeast-southwest flow) (*youngest*).

Almost certainly there was also a flow of ice from beyond the coastline of

Cape Province, with ice moving from west to east at one time and also from southwest towards northeast. On the coast of Natal there are signs that the ice flowed onshore from a source area which lay to the east, and we have seen already that ice from South West Africa flowed westwards to affect the Parana Basin and other parts of Uruguay and Brazil. These ice-movement directions are extremely difficult to interpret in terms of the present-day distribution of southern hemisphere landmasses. But when we look at the map of the 'clustered' continents of Gondwanaland (page 157) things become much clearer. The glaciation of southern Africa can be interpreted with any confidence only when we see the present landmass in the context of its Permo-Carboniferous 'neighbours': southern South America, the Falkland Islands, West Antarctica, and Madagascar.

Glacial deposits of Permo-Carboniferous age have been known in Australia since 1859, and field investigations have gradually revealed that the glaciation affected the greater part of the continent. The glacial features which have been mapped include glacial troughs, *roches moutonnées*, striated rock pavements, tillites, glacio-marine horizons, dropstones, and scattered erratics. These features have been explained as the products of both alpine glaciation and glaciation by one or more icesheets of great extent over the lowlands of southern Australia.

The glaciation seems to have commenced more than 300 million years ago in the mountains of Tasmania and southeastern Australia, and it has been suggested that these mountains had been largely worn away by glacial and other processes by the end of the Carboniferous period. In the earliest part of the Permian period, icesheet glaciation commenced in both western and southeastern Australia, and from the widespread distribution of thick tillites it seems that the greater part of southern Australia was inundated by ice on at least three occasions. The effects of glaciation were felt at least as far north as northern Queensland, as shown by the distribution of erratics and dropstones in glacio-marine sediments. Certainly there must have been ice shelves and wide areas of floating sea ice with icebergs well beyond the northern edges of the icesheets themselves. By about 270 million years ago the last icesheet had disintegrated, but there are glacial traces in younger deposits which show that some ice persisted (perhaps in highland ice caps and icefields) until about 250 million years ago.

The ice which affected southern Australia seems to have come from the south, and if we look at the map of Permo-Carboniferous Gondwanaland we see that the main ice centre must have been located in Antarctica. For a long time it had been suspected that glacial deposits of this age occur in Antarctica, but it was not until 1960 that firm evidence of ancient glaciation was found. Tillites of Permo-Carboniferous age have now been discovered throughout the length of the Transantarctic Mountains, in the Ellesworth Range of western Antarctica, and even in the nunataks which project through the present icesheet surface in Dronning Maud Land and elsewhere. In southern Victoria Land the tillites seldom exceed 40m in thickness, but in the Queen Alexandra Range and the Horlick Mountains there are exposures of tillites up to 400m thick. In the Pensacola Mountains there is a tillite which is at least 600m thick, and in the Ellesworth Range the Whiteout Conglomerate (which is almost certainly glacial in origin) is up to 900m thick. These huge thicknesses of glacial deposits are

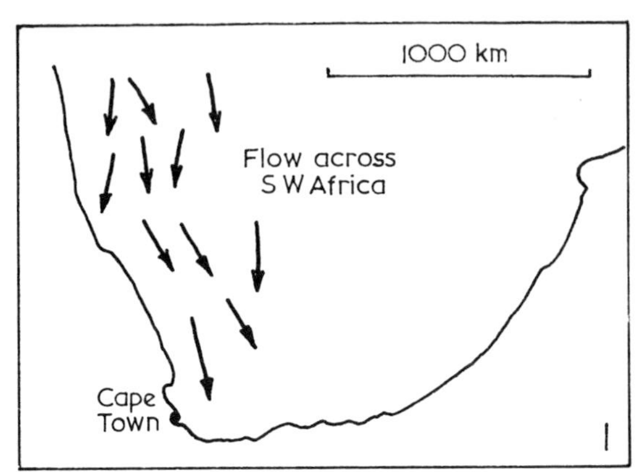

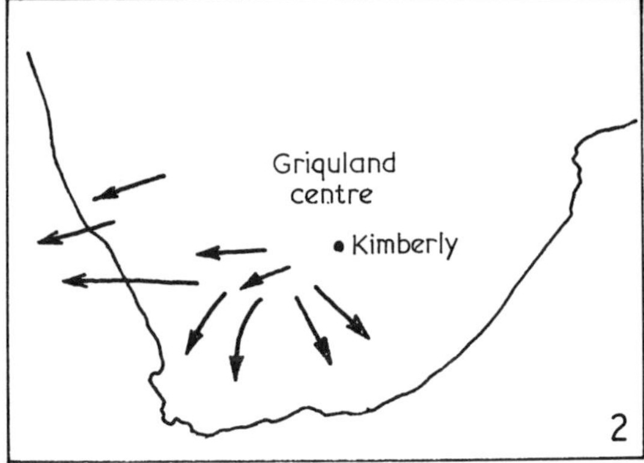

Directions of ice movement during phases of the Permo-Carboniferous glaciation of southern Africa. The 'onshore' movements of ice shown in map 4 may have occurred at about the same time as the ice outflow from Transvaal shown in map 3.

comparable to the thicknesses in South America and southern Africa, and just as in those areas they must indicate prolonged and repeated glaciation.

From the fossil remains of *Glossopteris* and other plants it seems that the glaciation of Antarctica lasted from about 290 million years until about 270 million years ago, and studies of the tillites in the Horlick Mountains show that there were at least five glacial/interglacial cycles. At present the evidence of ice-movement directions is confusing: in the Transantarctic Mountains the ice seems to have been moving from Victoria Land towards the Weddell Sea area, but there are also signs of ice movement both towards and away from the present south pole. Overall, the centre of the icesheet appears to have been in Wilkes Land; signs of a radial ice flow from this centre can be seen in the Transantarctic Mountains, Tasmania and southwestern and southeastern Australia. As more research is undertaken in the coastal districts of east Antarctica, and on various nunatak groups, the pattern of Permo-Carboniferous glaciation will gradually become clearer.

The Talchir glaciation of India was the first of the Permo-Carboniferous glaciations to be recognized, and the glacial evidence from central India was described as early as 1859. Now a wide variety of glacial features are known from many parts of India and Pakistan, including glaciated rock pavements, thick boulder beds, tillites, and glacio-marine horizons. In places the tillites are over 250m thick.

At the time of the ice age the northern margin of the subcontinent was part of the southern coastline of the Tethys, and this is the only area where 'non-terrestrial' glacial deposits have been found. There were highlands in the north-west, northeast and east, and these upland areas seem to have been the centres of glaciation and the main areas of glacial erosion. From these centres large ice caps sent outlet glaciers and piedmont lobes in all directions, leading to a wide-spread deposition of tills and fluvioglacial materials. Along the old northern coastline the glaciers affected a broad strip of lowland and at times flowed into the sea. The deposits in this area reveal a series of marine incursions which may be the result of interglacial rises of sea-level following the ends of specific

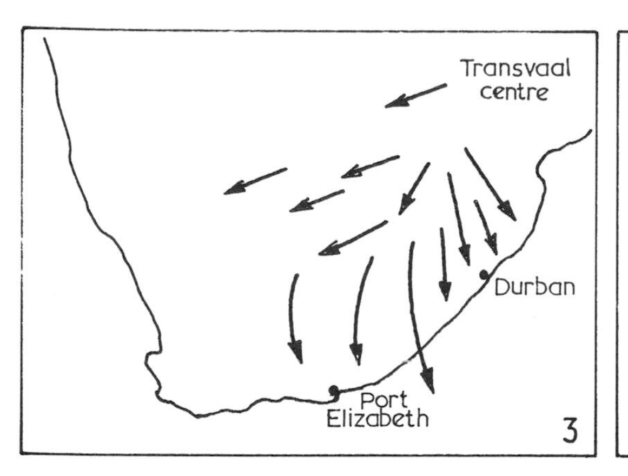

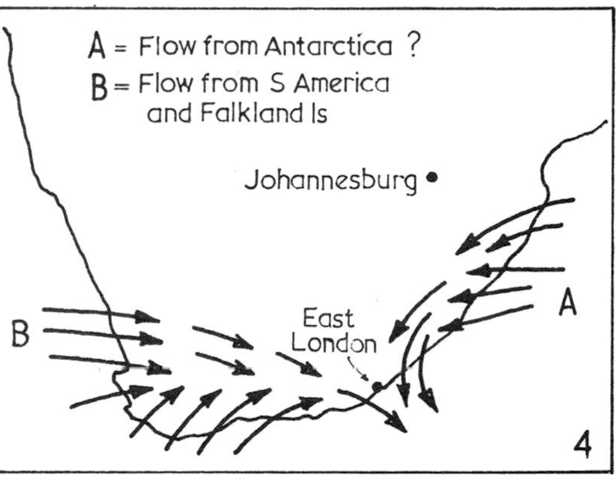

Smoothed and striated rock pavement at Irai, central India, clearly demonstrating the direction of local ice movement during the Permo-Carboniferous ice age.

glacial stages. However, the record is difficult to interpret with confidence.

The glaciation of India seems to have spanned the period between 310 and 270 million years ago, with the ice probably at its most extensive about 280 million years ago. Probably the main areas of glaciation so far described in the literature were areas of glacial *deposition*. However, there are signs of a northward flow of ice across the highlands of central India, and it is quite possible that the southern parts of the subcontinent were also affected by ice. Most likely this ice came from a radial outflow of ice from the main Gondwanaland icesheet. There were at least three major glacial episodes; and the final ice recession in the east and north, although accompanied by minor fluctuations, was very rapid. As indicated earlier in this book, 'catastrophic' ice disintegration seems to be typical of glacier margins in maritime situations.

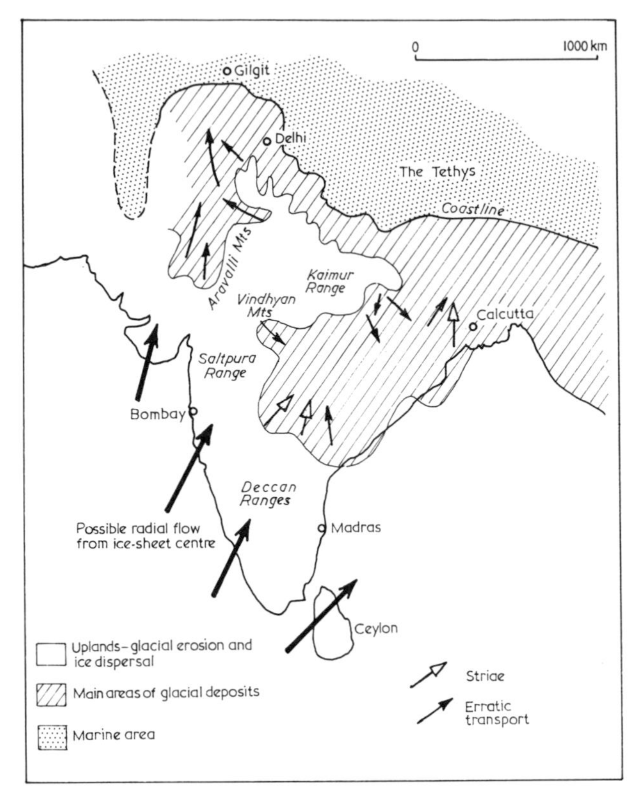

Some features of the Permo-Carboniferous glaciation of the Indian subcontinent. At this time Madagascar was located near western India and Antarctica lay to the east.

Traces of the ice age have been discovered from many other regions in both the southern and northern hemispheres.

In the southern hemisphere, for example, the Dwyka Tillite occurs in Madagascar, and glacial rocks considered to be of Permo-Carboniferous age occur well to the north of southern Africa in the Congo Basin and in Ethiopia. In the Falkland Islands there is a glacial sequence known as the Lafonian tillite at the base of the Gondwana rocks. In East Falkland it is up to 850m thick, and there is a different facies of tillite (which is, nevertheless, of broadly equivalent age) in West Falkland. The lithological characteristics and the associated fossils of these tillites show that they are quite clearly related to the glacial beds of both South America and southern Africa.

In the northern hemisphere there is a very localized stony 'mixtite' in Massachusetts which has been interpreted as a Permo-Carboniferous tillite. However, other evidence based on the sediment types and fossils of the Permo-Carboniferous period from the eastern USA show that the area lay well within the tropics at the time; if the deposit is really of glacial origin it must represent a very localized upland glacial episode.

Much more convincing evidence for glaciation comes from the northeastern part of Siberia. During the period in question this region lay close to the north pole, and so glacial or periglacial conditions were quite likely. The main deposits described within the last decade come from the Omolan River, east of the Verkhoyansk Mountains and northeast of the Sea of Okhotsk. The glacial horizons occur within a marine sequence, and their fossil content suggests an age of about 250 million years. Thus they appear to be somewhat younger than the glacial deposits of the main Gondwanaland ice age.

The Structure of the Ice Age

The foregoing paragraphs provide abundant evidence for icesheet glaciation on a vast scale across the high and middle latitudes of the Permo-Carboniferous world. Overall, the pattern of glaciation in Gondwanaland makes sense when one looks at the map of the 'clustered' landmasses, but it is still extremely difficult to unravel the main glacial and interglacial events which occurred around 300-280 million years ago. In places the glacial deposits have been eroded and redeposited; in places they have been buried by later deposits; and even where good sequences of tillites and glacio-marine beds are well preserved in the rock record it is often impossible to tell whether or not they give the 'whole' story of the ice age. In places there is evidence of ten or more glacial stages whereas elsewhere the record is so incomplete that only a single glacial stage can be identified with confidence. Also, the pattern of marine transgressions and regressions can almost never be correlated from continent to continent. Nevertheless, tentative conclusions, described in the next few paragraphs, can be reached concerning the distribution of ice and the events of the ice age.

The total timespan of the ice age was probably about 100 million years, although this includes a prolonged early period of glacier growth and a prolonged late period of ice wastage during which glaciers were probably restricted to the high-latitude areas and the high mountains. There is now strong evidence to show that the peak of the ice age, when huge icesheets repeatedly spread at least as far north as 35°S, occupied a period of about 40 million years between 310 and 270 million years ago. The greatest ice extent seems to have occurred around 280 million years ago.

During the ice age the central parts of the glaciated area were deeply inundated by a number of separate icesheets. Possibly these icesheets coalesced during the main glacial stages. In the peripheral areas (such as southeast Australia, northern India and the highlands of South America) there were large independent ice caps and highland icefields with outlet glaciers and ice piedmonts on the adjacent lowlands.

As Gondwanaland slowly moved with respect to the south pole the main centres of glaciation shifted accordingly. The palaeomagnetic evidence and the fossil remains from the tillites and their related sediments reveal the following sequence of ice centres: *earliest*—South America and South West Africa; *later*—South Africa, Madagascar and India; *latest*—Antarctica and Australia. There is little doubt that southern Africa was the focus of glaciation for the greatest length of time, and it is here that we find the most convincing evidence for substantial glacial erosion.

Along most of the margins of the Gondwanaland icesheet the ice edge was situated at sea. We have seen in the foregoing paragraphs that glacio-marine sediments are common in all the main regions affected by the Permo-Carboniferous icesheets. Even where glacio-marine sediments are rare or absent there are

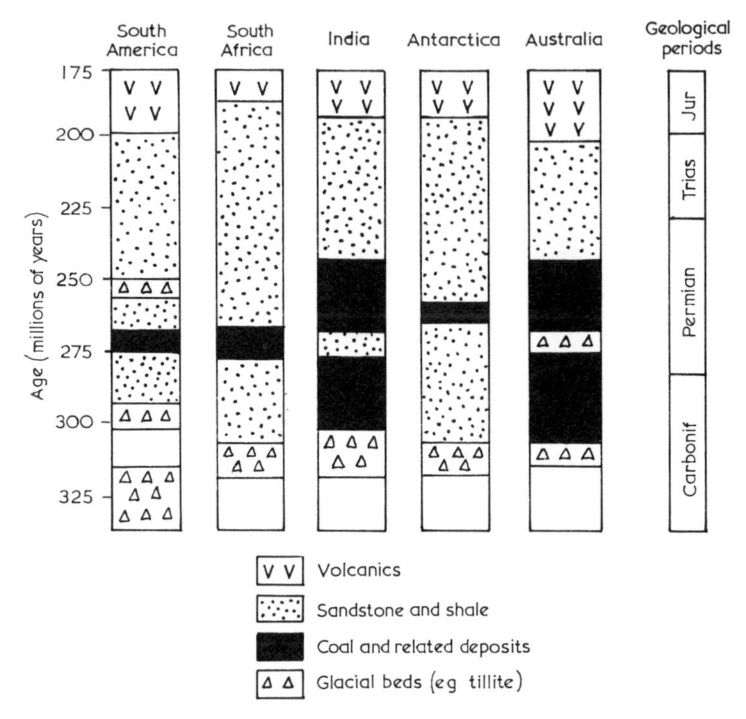

Columns showing, in a very generalized way, the main glacial and related horizons in the rock sequences of the Gondwanaland landmasses. The main tillites were deposited around 300-280 million years ago.

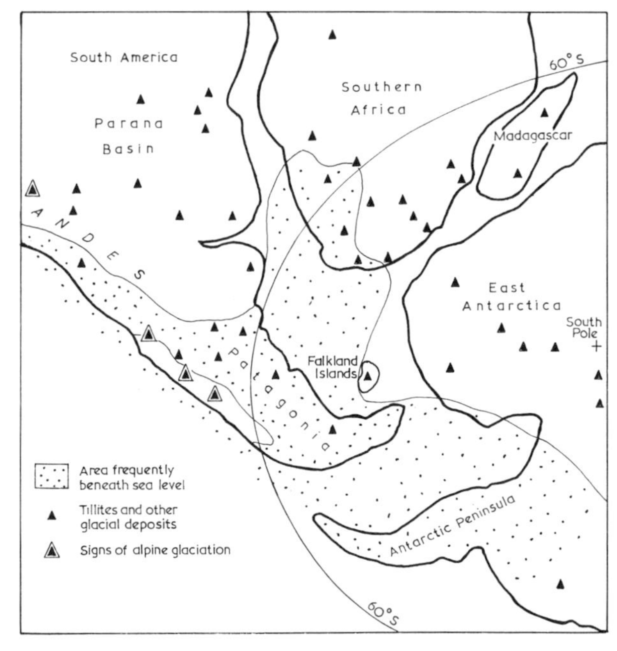

The possible location of the 'maritime' section of the Gondwanaland icesheet. Over most of the area shown on this map the icesheet bed lay near or beneath sea-level, making it very unstable.

interbedded tillites and marine shales and sandstones which show complicated relations between the sea and the ice edge. An icesheet or group of icesheets covering well over twice the area of the present-day Antarctic icesheet must have had a great influence on sea-level (see Chapter 3), and we can assume that the waxing and waning of the icesheets during glacial and interglacial stages must have been accompanied by sea-level oscillations of at least 150m and possibly as much as 250m. There must also have been substantial isostatic uplift and down-warping, leading to yet more complicated sea-level movements around the coasts of the supercontinent.

It is quite likely that large sections of the Permo-Carboniferous icesheets were warm-based. The great thicknesses of tillite in many areas suggest that there were prolonged periods of effective glacial erosion accompanied by deposition on a vast scale. The tillites of southern Brazil and Uruguay, over 1000m thick in places, suggest that glacial processes may have been particularly intense in the South West Africa and Parana Basin sections of Gondwanaland, and indeed this is an area which seems, from other evidence, to have been partly beneath sea-level at the time of the ice age (as in each of the other ice ages, the analogy with present-day West Antarctica springs to mind). Here, perhaps even more than on the other icesheet margins, catastrophic icesheet disintegration and large-scale surging behaviour may have been quite normal features of the ice age.

If the idea of Permo-Carboniferous maritime icesheets is reliable, it follows that surges accompanied by rapid sea-level rises may have occurred frequently during the ice age. The area referred to in the previous paragraph is at least as large as the area of the present-day West Antarctic icesheet. It has been calculated that the rapid disintegration of this icesheet would result in a sea-level rise of about 5m. Possibly, therefore, similar sea-level rises would have accompanied particular Permo-Carboniferous icesheet surges. This brings us back to the points made at the beginning of this chapter concerning the cyclothems in the Coal Measures of middle and late Carboniferous age. Glacial and interglacial sea-level oscillations of at least 150m and much more violent transgressions related to individual surges and phases of icesheet disintegration may well have been perfectly normal features of Permo-Carboniferous times. 'Glacial control' must surely be one of the more obvious explanations for not only cyclothems but also other changes in sedimentation in the shallow seas around all the coasts of Pangaea.

Ice, Plants and Animals

There is no doubt that the prolonged and intensive Permo-Carboniferous ice age was linked to a great cooling of the world's climate. In the tropics the climate was warm enough for tropical swamp vegetation to survive near sea-level and for deserts to form in areas remote from the coastline; but quite possibly there was a compression of the main climatic zones of both hemispheres as glaciers and permafrost gradually strengthened their hold upon the high and middle latitudes. We have already seen that large glaciers existed within 30° of the equator in early Permian times, and it is not unreasonable to imagine that parts of the tropics were also affected by permafrost.

As we noted earlier in this chapter, plant and animal life on land had to contend with climatic changes and also the elimination of many environmental 'niches',

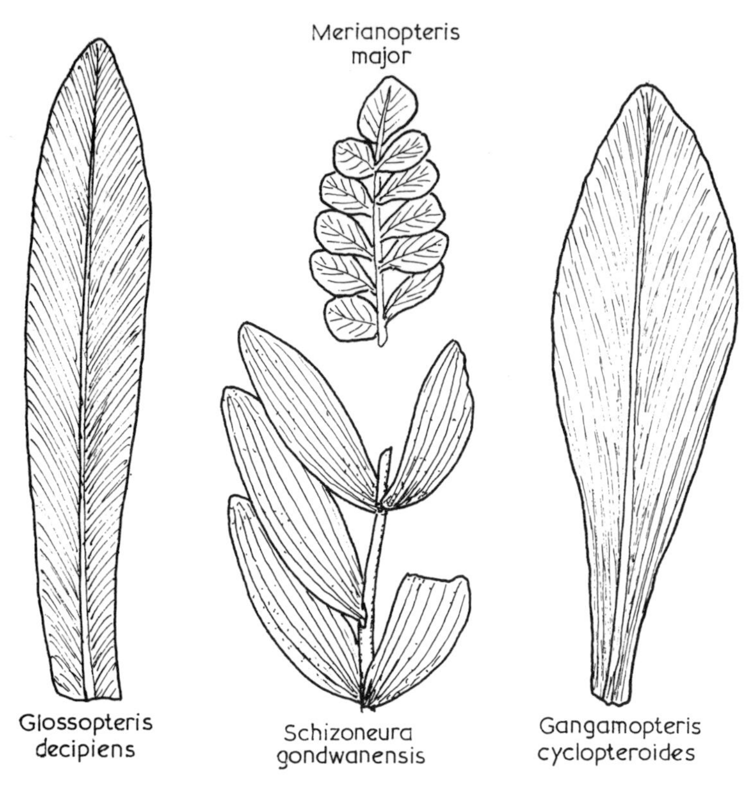

Some typical fossils from the *Glossopteris* flora of Gondwanaland. *Gangamopteris* probably evolved first, but the seed fern *Glossopteris* later colonized a vast area of Gondwanaland.

changing patterns of atmospheric circulation, and great competition within a few specially favoured regions. Glacier ice and permafrost, both waxing and waning over and again during many millions of years, placed great demands upon the resilience of many species which had evolved in the middle latitudes.

At sea, plants and animals had to contend with great temperature changes and changes in water cloudiness and salinity as meltwater and glacial sediments found their way intermittently into the oceans. With the 'clustering' of the continents into the vast landmass of Pangaea, ocean currents changed, altering water characteristics and disturbing the balance of many plant and animal communities. And the violent rises and falls of sea-level during glacial and inter-glacial stages led to the periodic flooding and draining of great areas of the continental shelf seas.

In view of these facts we can argue quite strongly that glaciation must have had many profound effects upon both plants and animals. But what precisely were these effects? How many species were really stimulated to adapt or evolve for success in new environments? How many species failed to adapt and so became extinct?

The evidence for glacial control over plant life is perhaps strongest in the case of the *Glossopteris* flora of Gondwanaland. As mentioned on page 157, *Glossopteris* was a genus of seed fern that grew over a wide area within and beyond the limits of the Permo-Carboniferous icesheets. There were a number of *Glossopteris* floras or plant associations, but palaeontologists have recognized for more than a century that all were peculiar to the landmasses of Gondwanaland. The principal members of the flora are easily recognized; they are preserved as fossils in rocks of many types, ranging from the coals of Antarctica and the terrestrial sandy beds of the Falklands to the sandstones and shales of Madagascar and the silty shales of the Indian Talchir Formation. Always the beds with plant remains are quite closely associated with tillites or other indicators of glacial conditions, and spores of *Glossopteris* have even been recovered from the tillites themselves.

This is what the palaeontologist Bernhard Kummel says about the *Glossopteris* flora: 'The striking individuality of the Gondwana flora can possibly be related to the cooling of the southern hemisphere during the late Carboniferous and to the widespread glaciation. The cosmopolitan flora that had occupied the area became extinct with the change to a cold-temperate climate, to which the Gondwana flora was adapted.'

Other authors have also commented on the growth of this unique flora in response to the climatic stimulus of glaciation, and E. Plumstead has suggested that *Glossopteris* and other plants evolved as a result of high cosmic radiation in the south polar regions. This is a rather 'special' hypothesis, and it seems more likely that the plants evolved (perhaps quite rapidly in geological terms) to occupy areas of permafrost and cool temperate conditions where prolonged winter freezes and short summers were facts of life. Since the first land plants had not appeared until Silurian times, the species associated with *Glossopteris* were the first ones which had to contend with extreme cold and violent seasonal extremes of temperature. Their widespread distribution is a testimony to their success in adapting to the climate of an ice age.

During Carboniferous and Permian times there were a number of important evolutionary events in the history of the world's plant and animal life, but by far the most dramatic period from the point of view of the palaeontologist is the later part of the Permian. Within a few million years there were drastic reductions in the number of genera of marine invertebrates: foraminifera, corals, ostracods, nautiloids, brachiopods and crinoids, trilobites, blastoids, three-quarters of the bryozoan groups, several of the cephalopods and sponges, and many families of molluscs became extinct.

On land, plant life was not greatly affected but vertebrates were. During the late Permian some 75% of amphibian and over 80% of reptile families disappeared. What was the cause of this 'biological catastrophe', this episode which has been referred to as 'the most spectacular extinction period in the fossil record'?

There is no agreement about the answer to this question. However, it is known that the various extinctions of plants and animals were not 'geologically instantaneous'; in other words, there was no single catastrophic event which wiped out large numbers of vulnerable species. It seems most reasonable to assume that the cumulative effects of the climatic and oceanic changes of the Permo-Carboniferous ice age, combined with sea-level fluctuations and the isostatic movements of Gondwanaland in particular, were in the long run too difficult for many plants and animals to cope with. Most species were, after all, still rather primitive and closely adapted to particular marine or terrestrial environments. These environments changed over and over again, on hundreds of occasions, as the ice age ran its course, and only the hardiest of species could survive.

With respect to other ice ages we can perhaps refer to 'glacial stimulus' as a factor in evolution; but quite possibly the Permo-Carboniferous ice age was so prolonged and so profound in its global effects that 'glacial stress' became a ruling factor instead. For a large part of the living world that stress was too great to bear. Only the most resilient of species survived the Permian period, and they inherited from the ice age a greatly altered world.

BRIAN JOHN

7

The Present Ice Age: Cenozoic

The last major period of prolonged glaciation in the Earth's history may have started some 37 million years ago in Antarctica, and by 12 million years ago some of the high mountain ranges of Alaska were glaciated. These dates for the onset of glaciation run counter to the long-standing concept that the last major period of geological time—the Quaternary System—is synonymous with the 'Ice Age'. We will show later in this chapter why this old idea is no longer valid; but, if we are to understand the events that led to the glaciation and which still have an effect upon the Earth, we have to consider the changes in the world's geography between the end of the Cretaceous (65 million years ago) and the middle of the Tertiary—say 30 million years ago.

There are several themes that run through this chapter. One of the most pervasive is the argument that the 'past is the key to the future'. This is a rather substantial misquote of the old geological adage that 'the present is the key to the past'. However, geologists and geomorphologists (especially those who study the rocks and landforms associated with glaciers and other natural systems that respond to global climatic change) are currently trying to see if patterns of Cenozoic glaciations and interglaciations can be recognized. If distinct patterns do emerge from their studies, meteorologists and climatologists will be able to model, and hence possibly predict, the future climatic trends that we and future generations will experience. This task is no ivory-tower academic exercise: we are becoming conscious that we inhabit 'spaceship Earth', that food production has a limit, that energy supplies have limits, and that economic and human disasters are occurring year after year because of the vagaries of climate. Most meteorologists now agree that overall climates do change, and in this chapter the clear evidence for major global changes of climate will be apparent.

We are living in a relatively rare moment of time when viewed over the last one or two million years; we are living at a time when major icesheets cover only Antarctica (with an area of 14 million square kilometres) and Greenland (1.76 million square kilometres). As mentioned in the early chapters of this book, we call such periods interglacials. Deep-sea records suggest that these periods of globally favourable conditions represent only about 10% of the last one to two million years. On average, the interglacials have lasted 10,000 years—our present interglacial started between 14,000 and 10,000 years ago and some claim that the next glaciation actually started 5,000 years ago! Other researchers, however, predict a return to ice-age conditions at times in the future that vary from 2,000 to 50,000 years hence.

Obviously we must refine these predictions if the world community is to take any notice of an impending natural disaster of unparalleled dimensions. But the human race is in something of a cleft stick. Although the disaster of a new ice age is obvious, what about the reverse—an increase of global temperatures and hence the melting of the Antarctic and Greenland icesheets? The melted water from the icesheets would run into the oceans, and world sea-level would rise approximately 75 metres. For a rather disturbing exercise, take a good atlas and look at the number of cities and densely populated areas of the world that lie below the 75-metre contour. It is difficult to estimate the percentage of the world's population that would be affected but, considering countries around the North Sea Basin, cities such as New York, Los Angeles, London, and Montreal, and large tracts of Asia, the impact of such a sea-level rise would be enormous.

This preamble explains why a knowledge of the Cenozoic Ice Age should really be part of the stock-in-trade of any intelligent person. This knowledge has deeper philosophical overtones also. From Chapter 2 of this book, it is clear that we do not yet understand what causes glaciations and interglacials. This is simply another way of saying that we do not understand how the world's atmosphere, oceans and cryosphere interact to cause our weather. Since this is the case, then it should be apparent that we do not know, and hence cannot predict, the likely outcome of *Homo sapiens*' unintentional and intentional modifications of our oceans and atmosphere by excessive CO_2 production, by agricultural practices which cause changes in the amount of energy that the Earth absorbs and re-emits to the atmosphere, and by other means.

In addition, the gradual cooling of the Earth during the early Cenozoic and the initiation of the current ice age have had a profound impact on the world's flora and fauna—all natural systems must be stressed by the repeated coolings and warmings that the Earth has experienced during this time. Some organisms have been opportunistic and have grown in species and numbers as less adaptable organisms have been wiped out by their inability to adapt to changing conditions. We have seen in the preceding chapters that such events have occurred many times in the geological record, but the current ice age is significant because our own species may have evolved as a result of the stimulus which it has provided. Of course, we will never really know if Man's primitive ancestors developed their successful living strategies solely in response to climatically induced stresses or whether a similar evolution would have occurred in a more placid world. At least it is fun to speculate.

Geological Nomenclature and Practice

For the last few million years of Earth history we have more data than for any other period. This is, however, a two-edged sword because the complexities of the local geological records inhibits the broad synthesizing studies that are required to erect models or hypotheses. For the Cenozoic, the geologists have recognized a multitude of different geological records, or types of records. These are organized into three formal stratigraphic 'codes', as follows.

Rock- or lithostratigraphy. This is the fundamental building block in all our studies. Lithostratigraphy consists of recording the succession of consolidated and unconsolidated rocks exposed in a river-cut, sea cliff, motorway embankment, quarry or drill hole and the naming of these units. At each section the evolution of the environment through time is evaluated according to the changes in the character of the sediments. Type-sections are set up so that other workers can examine the rocks and the succession and then make correlations between the type-section and the areas in which they themselves are working. The basic units of lithostratigraphy are known as Formations, subdivided into Members and Beds.

Biostratigraphy. In the older part of the geological record 'time' is told by the evolution of specific fossil groups; however, little evolution has occurred over the last two million years and hence Quaternary geologists use biostratigraphy more in order to deduce changes of climate than to tell the precise 'time' a unit was laid down. In both land and deep-sea sediments the main emphasis is now on changes in the *assemblages* of plants or animals. Thus biostratigraphy is that facet of geology which in part enables us to recognize past glacial stages and interglacials.

Time or chronostratigraphy. Chronostratigraphy is of fundamental concern to us. The Quaternary System is a chronostratigraphic unit and hence, by definition, the boundary between the Quaternary and the Tertiary, or between the Pleistocene and the Holocene, is intended to be *globally synchronous:* that is, the boundary represents the same age all over the world.

There are lengthy debates in the literature on the dates of these boundaries, and different workers assign dates of between 3 and 0.7 million years for the former and 14,000 to 7,000 years for the latter. However, these arguments miss the point. The arguments are all concerned with the differences in timing of global response to climatic change that is embodied in the concepts of the Pleisto-

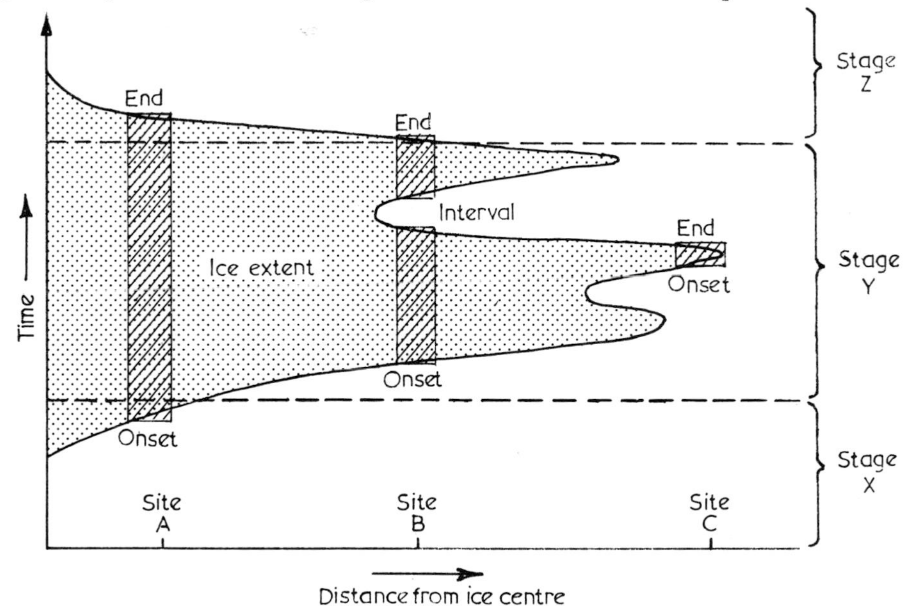

Idealized diagram to show how the divisions of geologic-climatic stratigraphy are related to time. At site A the first glacial deposits are laid down during stage X, and there is continuous glaciation for a great length of time. At site B the onset of glaciation is later, and there is an interval between two glacial 'stadia'. At site C glacial deposits are laid down during a very brief period in the middle of stage Y.

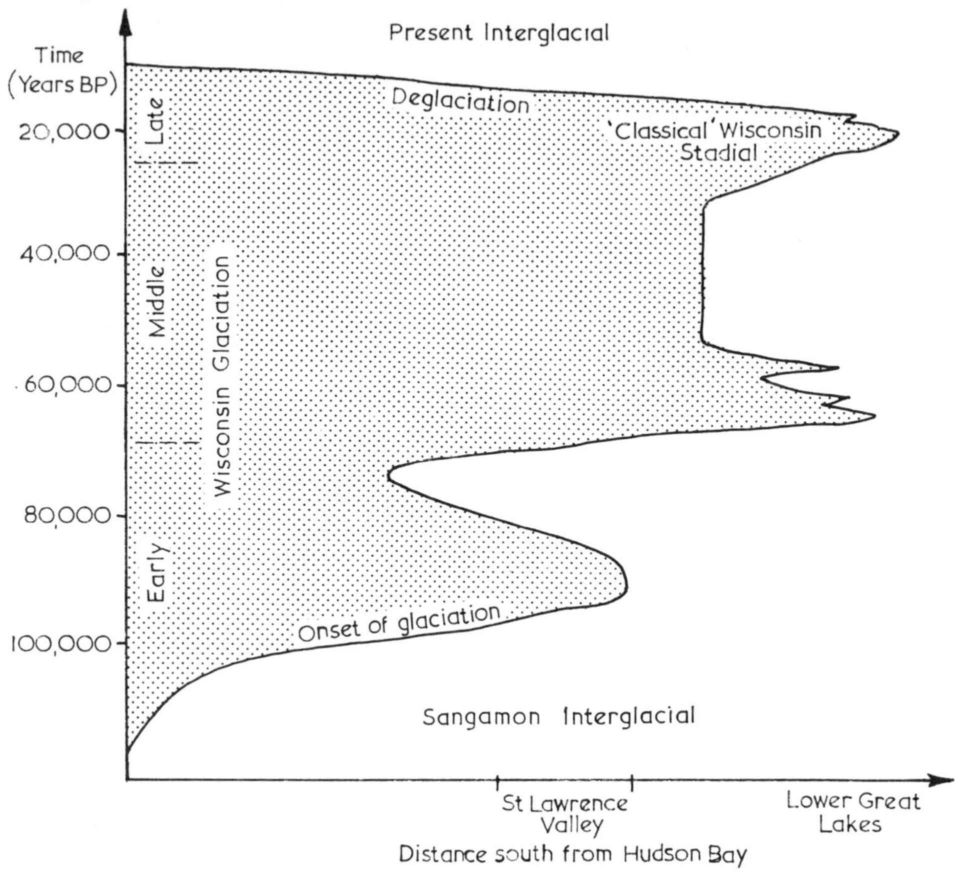

Time-distance diagram for the interpretation of stratigraphy along a line from Hudson Bay to the southern margin of the Laurentide icesheet. The diagram shows the history of glaciation during the Wisconsin glacial stage.

cene and Holocene Series. Obviously, the effect of a global cooling is felt first in the high latitudes and last in the tropics—but this does not mean that the chronostratigraphic boundary represents different ages depending on where you happen to be in the world. What it does mean is that, somewhere, a type-section has been designated and internationally agreed upon. Thus the type-section of the Pliocene/Pleistocene is in Calabria, Italy, where uplifted marine sediments record an influx of cool-water molluscs into the Mediterranean about 1.7 million years ago. It matters not one iota that cold-water molluscs reached the North Sea 2.5 million years ago—by definition the North Sea influx occurred during the Pliocene. These may seem like parochial and semantic problems, but they are worth noting because they cause confusion and poor communication within the group of scientists who work on problems of Quaternary climatic change.

Although they are greatly assisted by the above stratigraphic codes, geologists interested in the current ice age are aware of one other great problem of interpretation. It should be apparent from the above diagram that the onset of glaciation in the mountains and plateaux of Scandinavia or the eastern Canadian Arctic preceded the movement of ice across the Baltic or Great Lakes by several thousands of years. Thus the sediments which mark the onset of the last or

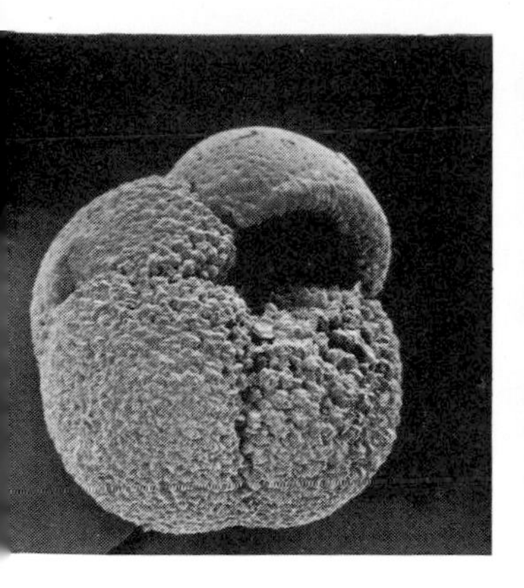 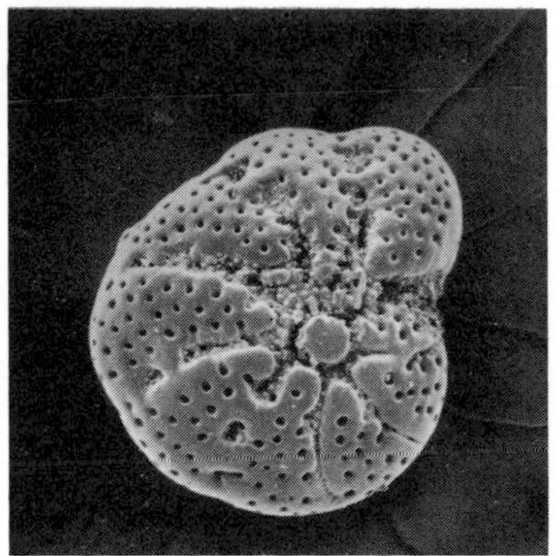 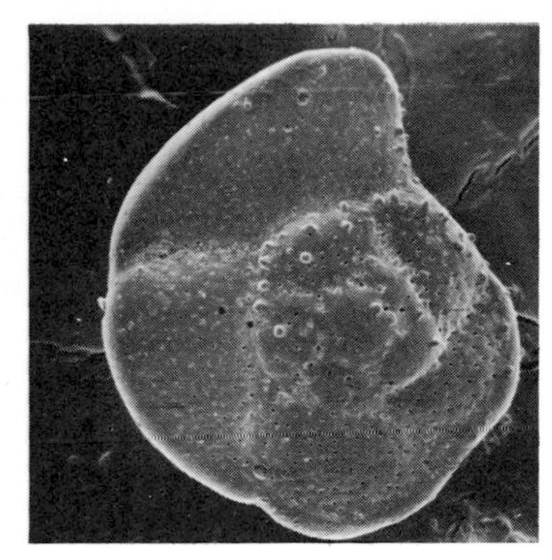

Foraminifera from deep-sea sediments.

Wisconsin Glaciation in Hudson Bay are from a notably different time than are the sediments in Illinois which represent the same climatic response. On the other hand, because of differences in ice dynamics, the sediment which represents the last glaciation in Illinois might be significantly thicker than near the ice centre in Hudson Bay. However, when the sites are correlated on a time-distance diagram the glacial (climatic) history of the area becomes more intelligible. The message in all this is that the *response* of natural systems, like icesheets, does not fit into tidy chronostratigraphic packages—an early glacial stage in one area may correlate with another region's late interglacial. Unless these points are understood the cause-and-effect relations of climatic change and glaciation/interglaciation are difficult to unscramble.

Dating of Cenozoic Rocks and Sediments

Geologists, physicists and chemists have been assiduous in their exploration of methods by which rocks, sediments and fossils can be dated. The complexities of the lithostratigraphic record for Cenozoic glacial and interglacial stages are impressive, and on land many sections have 1,000, 10,000, or even 1,000,000 years of missing record. How does one correlate a till in Spitsbergen with one in East Greenland if the intent is to try to build a picture of the global pattern of glaciation? Answers have been sought through two approaches. Firstly, methods have been developed to date relatively or 'absolutely' a sequence of geological events; and, secondly, attention has in large measure switched from probing into the records of glaciations and interglaciations on land to examining, in great detail, the fluctuations of assemblages of foraminifera and other organisms in long cores of sediment from the bottom of the deep-ocean basins. These sediment cores provide a continuous stratigraphic record for times that vary, for individual cores, from 30,000 to 3 million years.

There is no space here to attempt a catalogue of the respective merits and drawbacks of all the dating techniques currently available. Rather it is worth concentrating our attention on some specific methods that seem to be most important in studies aimed at understanding Cenozoic glaciations, especially the last few glacial/interglacial cycles.

A global picture of the waxing and waning of our Earth's icesheets is being obtained by studies of the relative changes of the heavy stable isotope of oxygen, O^{18}, in the shells of foraminifera extracted from deep-sea sediment cores. The steps required to end up with a diagram such as the one shown opposite are involved and not devoid of problems.

The real basis for the diagram lies in two separate areas: namely, the method of dating volcanic rocks called the Potassium-Argon (K-Ar) method; and, secondly, the study of the systematic variations of the Earth's magnetic field, palaeomagnetism.

The K-Ar method of dating is based on the assumption that volcanic minerals form without any contained argon-40 and that the argon-40 produced in the mineral by the decay of radioactive potassium is retained and can be distinguished from argon-40 in the atmosphere.

In the 1960s, workers combined K-Ar dates on volcanic rocks with studies of the palaeomagnetic field. We have already seen (page 40) that during the Cenozoic the Earth's magnetic field has several times changed from its present (or normal) direction to the reverse direction—and, of course, back again. For example, the last 700,000 years have been dominantly normal, although one or two short-lived reversals may have taken place. Prior to the 'Brunhes Normal Magnetic Epoch' the Earth's field was more inclined to be negative in direction, although there were several important reversals during the 'Matuyama Reverse Magnetic Epoch'. K-Ar dating demonstrated that the boundary between these epochs occurred about 700,000 years ago.

How does all this relate to the diagram above? The next step in the scientific chain was the recognition that iron-containing rocks in the marine sediment should preserve the magnetic signature of the time when they were deposited. Indeed, analysis of several cores in areas where low sedimentation rates were expected indicated that a long period of normal magnetic inclination and declination was followed by a reversal. The boundary is correlated with the Brunhes/Matuyama boundary dated in volcanic rocks, and so this provides a time-marker in the cores.

The influx of sediment into the deep-sea basins has been relatively uniform throughout the Quaternary and thus it is possible to use the 700,000-year date for the palaeomagnetic boundary and compute a sedimentation rate for the core. For example, if the 700,000-year boundary occurs in a core at 6.5 metres depth, the sedimentation rate works out as 0.00093cm per year or 0.93cm per 1,000 years; and thus the age of the core 1 metre below the core top is predicted to be about 107,500 years old. Predicted ages in the cores less than 40,000 years old may be checked by the radiocarbon dating of organic or carbonate materials in the cores.

We now have a core in which an age can be ascribed to a point in the sediment at any given depth. The next step is to process sediment from the core and to pick out the small fossil bodies (tests) of foraminifera. Then we analyse each sample to find out the relative proportions of the stable isotopes O^{16} and O^{18} present in it.

The initial interest and experiments were concerned with determining that foraminifera and molluscs are organic systems which grow in equilibrium with sea water. Laboratory studies indicated that, as the temperatures of the brines were changed, so did the ratio between O^{16} and O^{18}. Thus changes in O^{18}

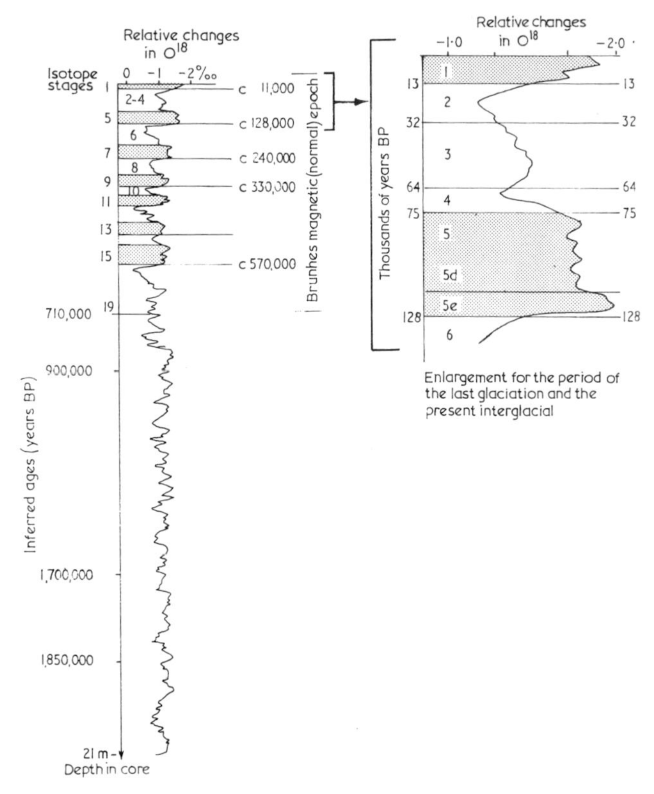

The record of variations in the O^{18} content of planktonic foraminifera from deep-sea cores. The areas of inferred interglacial conditions (by analogy with the present interglacial) are shaded.

could be related to temperature changes of the oceans. However, some researchers questioned whether the changes were really indicative of ocean temperatures. Rather, they suggested, the variations in the O^{16}/O^{18} ratio provide a faithful record of global changes in *icesheet volume*. In theory, during a glacial stage the lighter isotope O^{16} evaporates more readily from the oceans and is 'locked into' the great icesheets. This means that the proportion of the heavier isotope O^{18} left in the sea water increases. In contrast, during an interglacial stage 'light' water is returned to the oceans, so that the proportion of O^{18} decreases (i.e., the proportion of O^{16} increases).

If this theory is correct—and we should be aware that it may not be—then the oscillations on the diagram above may represent an important, indeed vital, record of icesheet fluctuations over the last several hundred thousand years. The peaks and troughs of the deep-sea O^{18} record have been numbered consecutively from 1, the last 10,000 years or so, to 23, which occurred close to 900,000 years ago. The peaks (1, 3, 5, etc.) are interpreted as interglacials or major interstadials, whereas the troughs (2, 4, 6, etc.) are interpreted as major glaciations. We will return to the implications of this work later in this chapter.

This outline of how deep-sea cores are dated and analysed is important because it emphasizes that one of the recent revolutions in the Earth sciences has been the increasing rôle of deep-sea studies as a unifying agent in contributing to our knowledge of Cenozoic glaciations. Indeed, many people are now firmly con-

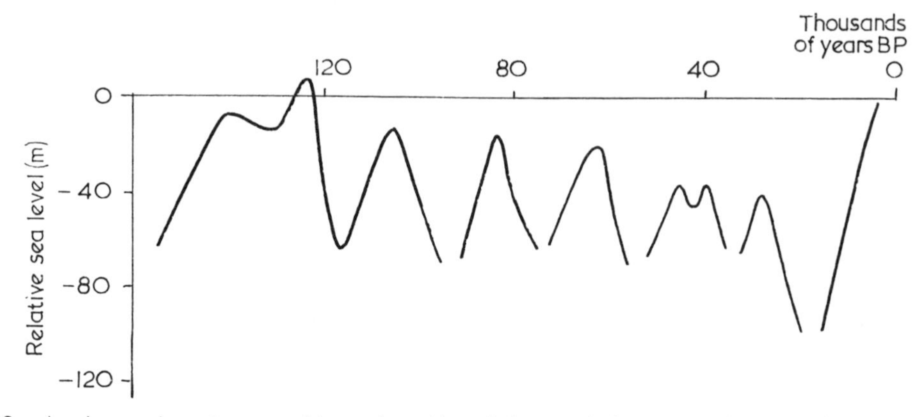

Sea-level curve based upon evidence from New Guinea and elsewhere where the effects of glaciation have been minimal. The timescale is based largely upon the uranium-series dating of corals in old marine terraces.

vinced that a global chronostratigraphy for the Cenozoic Ice Age must be based on the deep-sea O^{18} record.

However, the majority of scientists interested in Cenozoic glaciations work on land. They are thus faced with dating a stratigraphy that is more complex than that found in the oceans, and one which contains many missing strata. However, we should always be aware of the underlying assumptions in oxygen-isotope studies, and hence it is important to try to establish independently dated chronologies either in areas which have been directly glaciated, or in areas which record global changes of sea-level, as these are linked (not in a simple manner) with the growth and decay of the world's ice masses. Two dating techniques have been applied very successfully to such studies, although in the case of the radiocarbon (C^{14}) method the upper age limit is only between 40,000 and 75,000 years, whereas with the uranium-series (Th^{230}) dating method the age limit can be pushed back to 250,000 to 350,000 years ago. In the former case the timespan covers about one-half of the last glacial/interglacial cycle, while the Th^{230} range includes about five glacial/interglacial O^{18} variations.

Radiocarbon dates can be obtained on a wide variety of substances; for example, shells, wood, peat, bone and calcium carbonate precipitates. The method is based on the presence of small quantities of a radioactive isotope of carbon, C^{14}, in the atmosphere. This radioactive isotope enters the food chain, and all living things contain a known amount of C^{14} relative to the other stable isotopes of carbon, C^{12} and C^{13}. On the organism's death the C^{14} decays to form nitrogen at a fixed rate, so that every 5,700 years one-half of the radioactive carbon disappears. Sensitive counters have been developed which allow samples to be dated which are about thirteen half-lives (i.e., 75,000 years) old. At that age only 0.006% of the original C^{14} remains—of course, results must be interpreted with carbon, since such a small quantity might easily be contaminated, particularly by younger C^{14} which might have entered the sample by percolation of groundwaters or by mechanical infiltration such as the probing of the remains by plant roots.

In conventional C^{14} laboratories the counting apparatus actually measures

A log of driftwood stranded well above sea-level on a raised beach in the South Shetland Islands, Antarctica. Such materials can be used for the C14 dating of old sea-levels.

the rate of decay of the radioactive isotope; that is, it counts the individual pulses of energy given off as C^{14} atoms decay into atoms of stable nitrogen, N^{14}. For relatively young samples this does not take a great deal of time, but for older samples the rate of decay is so slow that samples are often measured over periods of days or even weeks. The actual decay process is random, and the number of emissions per unit time will vary. It is this variation which enters into the error term, such as in the date 5,340 ± 200. A more precise date (smaller ±) can be obtained by counting emissions from the sample over longer periods.

Recently, the prospects for increasing the efficiency, accuracy, and precision of C^{14} dates have improved enormously. Scientists in the USA and Canada have used linear accelerators and cyclotrons as mass spectrometers and have actually *counted* the number of C^{14} atoms in a sample. There are several advantages to this method: one is that the actual size of the sample counted is measured in milligrams rather than the present range of *grams;* the second is that the counting need be over a period of only 90 to 400 minutes, a far shorter time than that required by conventional laboratories.

A third advantage, not yet acted upon, is that it is theoretically possible to push the age one can determine using this method up to 100,000 years. This means that many critical interstadial sites in Europe and elsewhere, which probably date from between 75,000 and 100,000 years ago, can now be dated. However, it should be noted that samples in this age range will have few C^{14}

atoms, and so it will be imperative to ensure that they are not contaminated. This task will be difficult to accomplish, but it may be possible by concentrating the large-molecule proteins that are preserved in certain materials, such as marine shells.

Our knowledge of the association between O^{18} in the deep oceans and changes in global sea-level at the time of the last interglacial/glacial transition, between 125,000 and 115,000 years ago, rests mainly on the chronologies based on uranium-series dating of marine corals. In recent years this has been supplemented by the uranium-series dating of carbonate deposits in limestone caves from both North America and the UK.

The most common method used is called the ionium-deficiency method. Investigations have shown that modern marine corals, calcium carbonate precipitates in caves and, to a lesser extent, shells are formed with several parts per million of uranium but with negligible thorium-230: this is because uranium is soluble in natural waters whereas the product of its radioactive decay, Th^{230}, is insoluble. Through time the Th^{230} grows towards equilibrium with the uranium. The half-life of the process is 75,000 years. In contrast to C^{14} dating, where most laboratories can obtain dates of over seven half-lives (40,000 years or 0.8% of the C^{14} left), the effective upper limit for Th^{230} dates is 300,000 years or four half-lives.

A relatively new method of dating—or, more specifically, of correlating deposits—is based on the chemical changes that occur in the protein of a dead organism. Amino-acid dating is being applied to marine and land shells, bone, foraminifera, ostracods and plants. The principle behind the method is that, with time, there is a change in the optical configuration of amino acids held in the protein structure of the fossil. However, this is a chemical reaction and it is sensitive to temperature. Thus different results will be obtained for shell species of the same age if the samples come from say, southern England on the one hand and northern Alaska on the other. This would suggest that amino-acid dating methods will be most useful for regional correlations (for example, within the UK). With some knowledge about the functionings of the process the future may hold promise of wider applications. One of the most useful aspects of the method is that it has a very extensive time range; it can be used in nearly all areas on deposits up to one million years old, whereas in Arctic and Antarctic areas the technique can be applied to fossils that are up to fifteen million years old.

The dating methods referred to above depend completely upon the accuracy and reliability of the field stratigraphic observations made during the collection of samples for dating. There are a number of important questions. For example, are the materials to be dated *in situ*? Are the samples free of contamination? Are the origins of the strata above and below the material to be dated known for certain? In the long run our understanding of the Cenozoic ice age will come from a more sophisticated use of dating methods, and from continued efforts to understand the sedimentary characteristics and sequences of glacial deposits. With these points in mind we now turn to the large question of the development of the Cenozoic ice age.

The Development of an Ice Age

There are two basic questions to be raised in any discussion of the Cenozoic ice age. First, why and when did the large stable icesheets of Antarctica and Greenland develop? Second, what influence did this have on the development of unstable icesheets in the northern hemisphere?

The onset of Antarctic glaciation ushered in the Cenozoic ice age. It had been preceded by a long interval dominated by a mild climate and extensive seas, which led to the deposition of chalk in England and France. In North America, a large marine strait extended from Arctic Canada southward to the Gulf of Mexico. However, the history of the Cretaceous seas is one of repetitive marine transgressions across the land and regressions from it that are attributed to changes in the volume of the world oceans. In many ways, these variations of sea-level hint at a world which rather rapidly changed its geography during the early and middle Cenozoic. As mentioned in Chapter 2, these fundamental changes in the geography of our planet seem to provide the necessary condition for the initiation of an ice age.

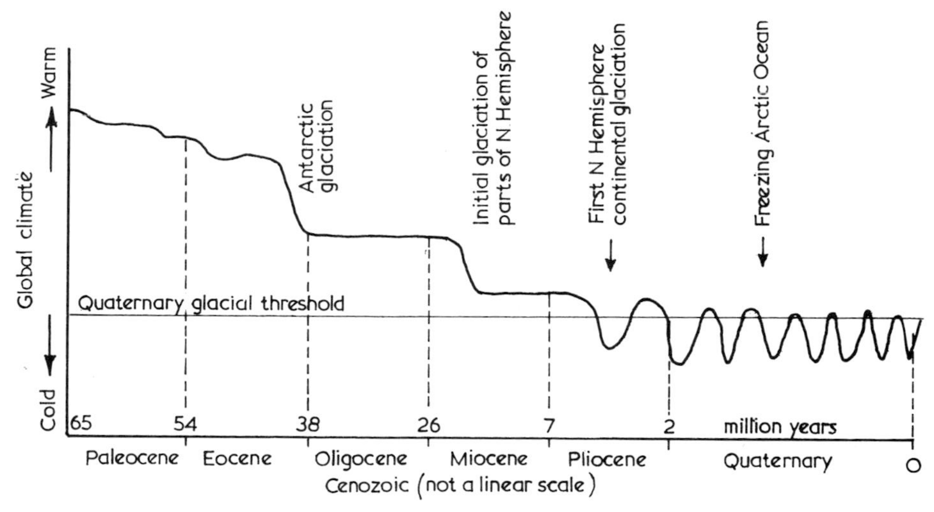

Suggested trend of global climate during the Tertiary and Quaternary towards a 'glacial threshold'. For at least the last 2 million years the climate has oscillated about this threshold.

The trend of world-wide temperatures over the last 70 million years is shown in the diagram. During the first half of the Tertiary, areas now glaciated, such as Baffin Island at 71°N, had vegetation and, by inference, a climate not unlike that of the USA's southern Atlantic coastal states. Similar conclusions have been drawn from studies of fossil plants and animals from sites in Alaska and northern Canada. These warm climates prevailed despite the fact that the north pole was located close to the edge of the Arctic Basin. Thus areas within the Arctic Circle of the time would have experienced long summer days and long winter nights.

However, the world was not static. Several things were happening which were to lead inexorably to the decline of global temperatures until conditions suitable for icesheet growth were prevalent. In the south, the continent of Antarctica was slowly drifting to the vicinity of the south pole. In the northern hemisphere, great palaeogeographic changes occurred, with the development of the Arctic Basin, the opening of the northwest (Baffin Bay/Labrador Sea) and northeast (Greenland and Norwegian Seas) extensions of the North Atlantic by spreading from the North Atlantic Ridge, and the active rifting between West Greenland and the eastern Canadian Arctic. In turn, the rifting of the old shields resulted in uplift and faulting of the rift margins. Uplift thus changed the topography in these northern areas from low plains into high uplifted rims and inland plateaux of moderate elevation of the order of 600 metres. The events in the northern and southern hemispheres were occurring simultaneously, and thus it is difficult to say precisely which event 'triggered' the onset of Cenozoic glaciation. They all did and, without this peculiar combination of circumstances, possibly the next ten to thirty million years of our world's history might have been different.

One of the problems facing scientists in their attempts to understand the Cenozoic ice age is the often contradictory data which they collect. But isn't this true of our present world? Trees grow on the ice-covered glacial debris in Alaska; beetles indicative of warm conditions have been captured near Icelandic glaciers; and glaciers presently exist in areas whose summer temperatures can be anywhere from 7°C to −4°C. As glaciation developed during Cenozoic time we might expect to see such apparent contradictions preserved in the geological record. Indeed, the evidence for the date of the onset of glaciation of the Antarctic Continent varies between 55 and 25 million years ago, while evidence of nonglacial conditions extends between 70 and 15 million years ago. However, few of these records occur in the same locality and the obvious implication to be drawn is that the glaciation of Antarctica was not simultaneous, but that early glaciation of limited extent occurred on one part of the continent at the same time as warm conditions prevailed over other areas. We should not forget that today glaciers do exist, at heights of 5,500m or so, on mountains which actually lie on the equator itself.

In Antarctica, the early sites of glaciation were the Transantarctic and Gamburtsev Mountains. The weight of evidence points to some glaciation of the Antarctic Continent by Oligocene time—that is, 38 to 26 million years ago. Probably a full-scale icesheet covered Antarctica by 20 to 25 million years ago, at the start of the Miocene.

The spread of an ice cover across the 13 million square kilometres of Antarctica would have a major effect on global temperatures. In Chapter 2 we explained how the summer ice and snow surface would reflect back into the atmosphere the majority of the incoming short-wave radiation from the Sun. This is because the reflectivity or albedo of a fresh snow surface is about 80%, compared to about 20-40% for vegetated surfaces; in other words, only 20% of the incoming energy is absorbed in the former case compared to 80-60% in the latter case. The glaciation of Antarctica would therefore naturally result in a significant drop in average temperatures all over the world, and this would bring the world closer to the threshold of widespread glaciation.

The gradual deterioration of the world's climate during the early and middle Tertiary is clearly expressed in the changes of vegetation in northern latitudes. Through time, warm-temperate to subtropical plants were replaced in Alaska and northern Canada by colder-climate species. However, the first direct evid-

Glaciers in the Transantarctic Mountains of Antarctica. The earliest glaciers formed on the southern flank of the mountain range and flowed towards the south pole. Today the mountains are overflowed by ice from the East Antarctic icesheet, with outlet glaciers flowing away from the pole towards the Ross Ice Shelf.

ence of glaciation in the northern hemisphere occurs in the high mountains of southern Alaska where there are glacial deposits interbedded with volcanic lava flows. The latter can be dated by the K-Ar method, and the results indicate that these high mountains maintained an ice cover during middle Miocene times, about 10 million years ago.

One of the major unknowns at this moment is the date of the inception of the Greenland icesheet. No direct evidence has been found on Greenland itself. On general grounds, it seems that this icesheet may have grown during middle Miocene times. Here the deep-sea cores once again demonstrate their importance in studies of Cenozoic Earth history. In recent years a major exploration of our planet has been carried out by the Deep Sea Drilling Program (DSDP). DSDP is based on a ship, a rather special vessel, which is used as a drilling platform. Long cores have been taken from many of the world's oceans, and these may provide the answer to our question. In a core from a site south of Greenland and northeast of Newfoundland, the first evidence of detritus deposited from beneath floating ice occurs at a level in the core which is dated at 3.5 million years ago, or in the mid-Pliocene. Piston cores from the Arctic Ocean support this general figure as they indicate that the first evidence of such deposition into the Arctic Basin occurred some time between 3 and 6 million

(*Above*) Nunataks of Cenozoic basalt projecting through the surface of a small Icelandic ice cap. In Iceland the interbedding of volcanic rocks and glacial sediments indicates that the onset of glacial conditions occurred at least 3 million years ago.

years ago. Candidates for areas where early glaciation may have occurred around the shores of the Arctic Ocean would include Greenland, the Canadian Arctic Islands, Iceland and Svalbard. At the moment we know nothing about the onset of glaciation in any of these areas, apart from Iceland—where again the presence of volcanic rocks interbedded with glacial sediments suggests that there have been many glaciations during the last 3 million years.

We have seen in the last few paragraphs the evidence for the world's climate sliding inexorably toward the 'instability threshold' which has been characteristic of the last 1 to 2 million years. It appears that by 3 million years ago the Greenland and Antarctic icesheets were in existence. Mountain glaciation was a feature of parts of the Alaskan mountains, and the first local ice cap may have lain over the plateau basalts of Iceland. In other words, the world of 3 million years ago may have been rather similar to our present interglacial world. Evidence suggests one significant difference. Whereas the Arctic Ocean is today largely covered by sea ice, analyses of sediments and the fauna from deep-sea cores indicate that 3 million years ago it was ice-free. The present 'iced-in' character of the Arctic Basin is estimated to have commenced as recently as 700,000 years ago. The importance of this freezing-over as regards our world's climate cannot be overstressed. Although the surface of the sea ice in summer is not as reflective as glacial ice or snow, it is nevertheless significantly more so than open water (an open ocean absorbs most of the incoming short-wave radiation). Thus the advent of an ice cover on the 15 million square kilometre Arctic Ocean would have been comparable in its effects to the earlier inception of the Antarctic icesheet. The date for the freezing-over of the Arctic Ocean is

(*Opposite*) An east Greenland fjord full of icebergs. Icebergs such as these began to carry glacial debris into the seas around Greenland about 3.5 million years ago.

The frozen Arctic Ocean off the coast of northern Ellesmere Island. The Ward Hunt ice shelf (shown in this picture) and the surface layer of sea ice first formed about 700,000 years ago.

thus an important event in the history of the Cenozoic ice age. The date of 700,000 years ago is based on the palaeomagnetic timescale noted earlier, and indicates that during the entire Brunhes magnetic epoch the frozen polar ocean has been a major influence on world climate.

We now have two basic images: first a world about 3 million years ago with extensive icesheets in Antarctica and Greenland, some local ice here and there, but with an ice-free polar sea; second, a world about 700,000 years ago where during interglacial times ice extent was similar to that of the present. What happened between 3,000,000 and 700,000 years ago? What dates can be attached to the onset of the classical glacial sequences of North America and Europe—the former commencing with the Nebraskan Glaciation of the Great Plains of America and the latter in the European Alps with the Donau Glaciation? These are important but difficult questions. The type areas for these deposits contain no materials that can be directly dated at this time, although a reasonable estimate of the age of the Nebraskan Glaciation is possible because of association with volcanic ashes. However, in a chronological sense, the period around 700,000 years ago marked the time when the global climate dropped to the 'glacial threshold', and glaciation of the northern hemisphere by massive unstable icesheets began. This was a major turning point in the Cenozoic ice age. Although we cannot be absolutely certain as yet, the evidence suggests that Antarctica (certainly East Antarctica) and Greenland have maintained their ice cover for the last 3 million years. These are the stable icesheets of the world. Similarly, the Arctic Ocean has maintained its ice cover for the last 700,000 years, and it too represents a constant in the global energy balance. The variables in the last 3 million years of our Earth's history have been the unstable elements of the large northern hemisphere icesheets, the possible large extent of ice shelves fronting the icesheets, and extensive areas of sea ice. In Chapter 2, the causes of the onset of glaciations were discussed; later in this chapter we will turn to what many people regard as an even greater problem—namely the problem of what stops a glaciation once it has started.

The Pattern of Late Cenozoic Glaciations

We know that there were four major glaciations during the Cenozoic in Europe and North America: in the Alps these have been called (from oldest to youngest) the Günz, Mindel, Riss and Würm glaciations, whereas in North America the supposedly equivalent events are called the Nebraskan, Kansan, Illinois and Wisconsin glaciations. In the Alps, two earlier glaciations, the Biber and Danau, are recognized by some authorities. No matter, the picture we used to have of our late Cenozoic world was of one predominantly interglacial in nature but interrupted by four cataclysmic events, the glaciations, of unknown duration. The ever-expanding analysis of the deep-sea cores, combined with the ability to date an increasing range of materials, has brought the old and new into face-to-face combat.

How can the diagram on page 179 be squared with the old concept of four glaciations? The O^{18} record from deep-sea sediments suggests, if some of the basic assumptions are correct, that nine or ten times during the last 900,000 years the O^{18}/O^{16} ratio in the world's oceans has shifted to such an extent that the growth of the major northern-hemisphere icesheets is required to explain the data. Thus, on a global scale, the O^{18} evidence seems to suggest that, during the last 900,000 years or so, massive areas of the world were inundated by icesheets; these icesheets survived for approximately 90,000 years on average and then disappeared. If we take the O^{18} values of foraminifera of the last 10,000 years as indicating values typical of *interglacial* conditions, then one of the major conclusions of these studies, of direct significance to mankind, is that interglacial periods have been much shorter than glacials and have on average lasted only 10,000 years. In the longer timescale of the late Cenozoic, the deep-sea O^{18} record does suggest a division into two periods, an earlier period dated between 2,000,000 and 850,000 years ago on the one hand and the last 850,000 or so years on the other. This is a qualitative impression of the O^{18} records. The earlier period does not show the same large swings of the proportion of O^{18}. In the last 800,000 years or so the glacial/interglacial swing has been about 1.5% O^{18} content, whereas prior to that the record from a core in the central Pacific Ocean shows a swing of about 0.7%.

The last glaciation (Wisconsin in North America, Würm in the Alps) was not as extreme as the preceding several glaciations. In turn, this implies that the amount of water tied up in the world's icesheets during this last glaciation was not comparable with the other glaciations of the Brunhes magnetic epoch.

On a global scale, Broecker and Van Donk have drawn attention to the saw-toothed pattern of each glacial cycle. This indicates a slow, gradual build-up of ice to a maximum which is then followed rapidly by deglaciation. In this model, the rapid disappearance of the unstable icesheets contrasts strongly with the slow build-up phase.

The deep-sea O^{18} record suggests that there are recurring cycles in the world's response to climatic change. Analysis of the oxygen isotope record using statistical techniques suggests that there are cycles of about 20,000, 45,000 and 100,000 years that have the same frequency as changes in the distribution of incoming short-wave radiation from the Sun deduced from celestial mechanics (the so-called Milankovich theory, described in Chapter 2). This basic patterning of global response to climatic change should not blind us to the fact that we may expect quite large regional variations from the global pattern, and that the cycles discovered in the deep-sea record are averages.

How do we reconcile the traditional view, that has been developed by glacial geologists over the last century, of four major glaciations with the more violent interpretation of the Earth's late Cenozoic climate as shown by studies of deep-sea sediments? This is a fundamental question and is difficult to answer, primarily because so many of the classic sites upon which the glacial stratigraphy has been erected cannot be dated. A link between the continuous strip-chart record of the oceans and the instantaneous scattered readings of the land glacial record may be the history of climatic change preserved in the soils and land shells of the thick wind-blown dust deposits (loesses) of eastern Europe, particularly Czechoslovakia, Hungary, and Austria (see Chapter 3). These sediments do maintain their magnetic signature, and hence it has proved possible to distinguish sediments of Brunhes age from the older Matuyama deposits. George Kukla and other scientists have reconstructed the climatic history of central Europe back for about 1.8 million years. Over this period some eighteen 'glacial' cycles are recognized. These data have been interpreted to indicate that the amplitude of climatic change between 700,000 and 1,800,000 years ago and the last 700,000 years has been similar. This conclusion contradicts an earlier statement, and thus it is obvious that as yet no consensus is available on this topic. If there were eighteen glacial cycles, how do these tie in with the fourfold glacial sequences in Europe and North America?

A preliminary attempt to propose a solution to this question has been made by Kukla; it is concerned primarily with the last 900,000 years, and the earlier events have yet to be associated with the deep-sea record. Kukla's main contentions are as follows: that the classical interglacial sequences are frequently complexes which extend through several of the marine isotope glacial/interglacial stages; and that the glaciations older than the last one include a series of glacial episodes equal in magnitude to the last Wisconsin/Würm Glaciation. For example, in North America the last interglacial is named the Sangamon Interglacial; Kukla now suggests that deposits considered to be correlative with this interglacial include materials spanning the period 320,000 to 70,000 years ago. Similarly, the last interglacial in Northern Europe is frequently called the Eemian Interglacial; however, Kukla suggests that there were three episodes which have been called 'Eemian'—these are correlated with marine isotope stages 5, 7 and 9, respectively (see page 179).

The entire problem is really one of dating. If isolated sites are found where the strata include interglacial deposits older than 70,000 years (older than most C^{14} laboratories can handle), how can these be related one to another? The sites come from different areas and hence correlation on biostratigraphical grounds is not straightforward. If the interglacial stratum contains freshwater or land bivalves and snails, then the amino-acid method is one method by which it can be decided whether an 'Eemian' interglacial bed from Denmark is close in age to the supposed equivalent, the Ipswichian, in England. At present, the diagram above should be taken as a very stimulating hypothesis that requires testing wherever possible.

It is going to be very difficult *firmly* to fix the classical glacial and interglacial sites into the stratigraphy of the deep oceans. However, the number of glacial and interglacial cycles that are apparent in the O^{18} record from marine cores can be matched with the loess cycles. Furthermore, long cores in lakes and peat bogs from Europe and South America have been obtained that go back half a million years or so.

One core from Macedonia was twelve metres in length and extended back to isotope stage 17 (see page 179). The variation in oak pollen provides a measure of the local vegetation and hence regional climate: the oak percentages rise during interglacial periods and maintain low values during glaciations. The fluctuations in oak pollen indicate several major changes of climate that can be correlated with the deep-sea record from the North Atlantic with a minimum of fudging. Most of the core cannot be dated, of course, and hence the matching of the deep ocean and Macedonian records cannot be considered an absolute test.

It is, however, a strong test and one which suggests that Europe has suffered many more climatic fluctuations than were allowed for in the classical model of glaciation. The main difficulty lies in the correlation of events earlier than the last interglacial. Because it is so much closer to us, because we can date and understand some of its characteristics, and because the fauna and flora of our world owe a great deal to it, we will next discuss in more detail the events of the last 125,000 years.

The Last Interglacial/Glacial Cycle

This section deals with the sequence of events, as we know them, that indicate the speed of glaciation and deglaciation. In addition, we will look at the world, especially the northern hemisphere, at the height of the last glaciation. This maximum is not really particularly well dated, but by convention many consider it to have happened about 18,000 years ago. Again we rely for our global view of events on the deep-sea O^{18} variations.

The peak of the last interglacial occurred about 125,000 years ago. Evidence from around the world indicates that the last interglacial was considerably warmer than at any time during the 10,000 years or so of our current interglacial. For example, in the Canadian Arctic last-interglacial sites have been located and plants, mosses and beetles used to estimate the former climate by analogy with the present distribution of key species: larch grew on Banks Island 300km north of the present northern limit; dwarf birch was the dominant shrub in the Low Arctic tundra which covered Baffin Island, whereas today dwarf birch is

Evidence of the high sea-level of the last interglacial. An old raised beach of Eemian or Sangamon age on the coast of Wales. The beach pebbles rest on a rock platform which is probably much older.

only to be found 450km south of these sites, and even there it occurs only in especially favourable localities; and at the same sites on Baffin Island the mosses and beetles show distinct similarities to assemblages which occur today about 1,000km to the southwest near Hudson Bay. This warm last interglacial was marked by a world sea-level five to eight metres *above* that of the present. A five-to-eight-metre global rise in sea-level may not seem much, but it is equivalent to the mass of water stored in the Greenland or West Antarctic icesheets. The high sea-level indicates that one of these two icesheets probably disappeared about 125,000 years ago. The weight of evidence favours the view that it was the West Antarctic Icesheet which collapsed and disappeared.

Thus, 125,000 to 120,000 years ago, the geological evidence indicates that the world was rather warmer than at present and that less glacial ice existed. And yet, by 115,000 years ago, only 5,000 years or so later, the evidence from studies of sea-level change and the O^{18} record indicates that the world was halfway into the last glaciation!

This is a remarkable and rather disturbing conclusion. In 5,000 years, enough water was transferred from the oceans onto the land for world sea-level to drop by sixty metres. This averages out to a 1.2cm fall of sea-level per year, or twelve metres per thousand years. These values may not seem large or even extreme: to put them in perspective, they have to be seen against the immense area of the world ocean, which totals 360 million square kilometres. Thus, during the onset of the last glaciation, 43,200 cubic kilometres of water (equivalent to about 129,600 cubic kilometres of snow) were being stored on land each year. It is hard to get a perspective on such figures, but we should be impressed!

The onset of the last glaciation was, when seen in the context of geological time, a dramatic event. However, the speed of ice build-up poses its own special problems. In the cold areas of northernmost Canada annual snowfall equals only about 0.1m (in terms of water), and near the present treeline the values vary between 0.3 and 0.5m. Of course, most of this melts and disappears each summer. By contrast, polar areas which have a more equable maritime climate,

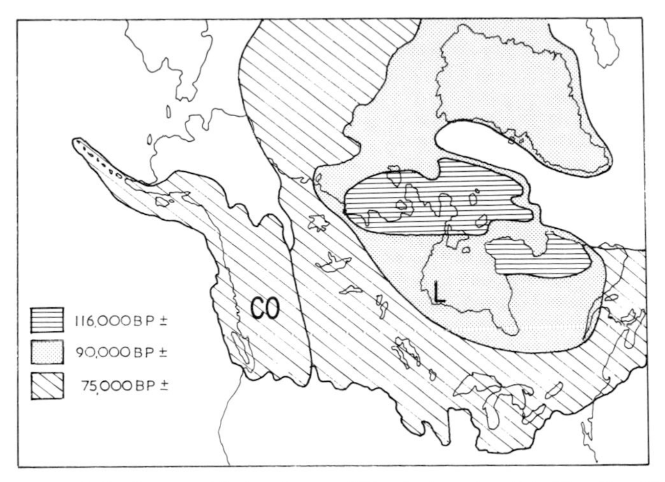

The three segments of the Laurentide icesheet of North America (L). The Cordilleran icesheet (CO) is also shown. The 'nucleating area' was ice-covered by 116,000 years ago; the stable core by 90,000 years ago; and much of the unstable periphery by 75,000 years ago.

such as West Greenland, Norway and Spitsbergen, have appreciably higher amounts of snowfall. The paradox of the last glacial inception is how to build icesheets at the rates suggested by the geological evidence. The rates are such that more moisture needs to be pumped into the areas of icesheet growth, and this in turn implies that in such areas temperatures may not have declined; they could even have risen.

The studies on sea-level have been carried out on subtropical islands; the deep-sea O^{18} records are also physically removed from glacial centres and thus we must ask whether there is any evidence in the polar areas for the sort of scenario sketched above?

The answer is yes. The major icesheet of North America was the Laurentide Icesheet. This vast icesheet, once the size of the present Antarctic Icesheet, covered the whole of central Canada and extended outwards to the coasts of Arctic Canada and Labrador and into the northern states of the USA. To the west it abutted at times against another icesheet located over the Rocky Mountains and coastal mountains of Canada, and to the north it was bounded by another complex of ice variously termed the Innuitian Icesheet or the Franklin Ice Complex. The Laurentide Icesheet was composed of three geographical segments: an area of initial ice accumulation; an area which contained a stable icesheet between 90,000 (?) and 9,000 years ago; and a wide southern and western margin where the evidence indicates that the icesheet behaved unstably over periods of thousands of years.

Baffin Island lies at the northeastern margin of the Laurentide Icesheet and its plateaux and mountains are today partly covered by ice caps and glaciers. The high rolling uplands of Baffin Island, north Labrador, Keewatin, and islands of the Canadian Arctic Archipelago are all considered to be areas where the last icesheet of North America became established. It is only 100km from the uplands of interior Baffin Island to the outer coast which fronts Baffin Bay and the Labrador Sea. Today the outer coast is bathed in cold arctic water carried southward from the Canadian Arctic Islands by the Canadian Current;

water temperatures are $-1°C$ and, of course, among the fauna in the sea and the flora on the land adjacent to this cold current only hardy species survive. One would think this area would be positively miserable during the onset of the last glaciation, but investigations show otherwise. Above the peaty sediments of the last interglacial are marine sediments which contain shells of organisms similar to those which today live in the warmer waters of the West Greenland Current and off Northern Norway. With the disappearance of the colossal weight of the overlying glacier, the land has risen, so that these marine sediments are found at elevations as high as 70m above present sea-level. Above the marine sediments is a glacial till. The marine shells have been dated by a number of methods and they indicate that the growth of an icesheet and the advance of its outlet glaciers occurred after 130,000 and before 86,000 years ago.

Our word-picture for the initial conditions which caused the glaciation of North America must include, therefore, an added influx of warm water from the West Greenland Current. This would cause more open water in Baffin Bay and probably an increase in snowfall over Baffin Island.

What was happening in Northern Europe at this time? Evidence is available from cores from the North Atlantic plus continuous sections on land. The Greenland and Norwegian Seas represent an enlarged image of the oceanographic situation in the Labrador Sea and Baffin Bay: warm water moves northeast as the North Atlantic Drift and leads to the mild maritime climates of the British Isles and Norway; cold arctic water flows southward along the coast of East Greenland. The deep-sea cores from the northern North Atlantic indicate that about 120,000 years ago the North Atlantic Drift was forced south of its present position by the movement of cold polar water. About 115,000 years ago the polar front (the oceanographic boundary between cold arctic water and warmer subarctic waters) extended in an east-west line from the vicinity of the British Isles. This happened twice during isotope stage 5; these global stadials are referred to as 5d and 5b. These sudden rather short pulses of cold are recorded in the sediments of North Europe. They are beautifully displayed in an 18m core from the Grand Pile in France.

Our picture of the northern hemisphere during the onset of the last glaciation includes, then, a more open Baffin Bay and Labrador Sea than at present, and conversely a much colder and more severe climate in Northern Europe, at least as far south as latitude 50°N. General causes of glaciation have already been enumerated in Chapter 2, but it is worth noting that, once the critical threshold had been reached, each glaciation may have been caused by a unique set of conditions, some of which may have involved mechanisms like those of the Milankovich theory, whereas others may have been random elements with no underlying strong periodicity in their occurrence.

During the period 120,000 to about 80,000 years ago there were two main glacial stadials on the global level. These were separated by two milder episodes when some of the accumulated ice melted and the world sea-level rose from about sixty metres to about fifteen metres less than today's level. Full glacial conditions occurred very rapidly close to 75,000 years ago. Before that we can surmise that the glaciation was limited to growth and expansion of the Antarctic and Greenland icesheets, the development of the stable core of the Laurentide Icesheet, the growth of ice masses in the Canadian Arctic Archipelago, and the development of an ice complex in the European Arctic, possibly centred on the shallow shelf of the Barents Sea. There is no evidence that northern Europe was glaciated during this early phase of glacial growth.

The rate of change in the O^{18} record and the rate at which world sea-level fell about 75,000 years ago matches that of the earlier event of 120,000-115,000 years ago. World sea-level fell to about 100m below its present value in a matter of a few thousand years.

At 75,000 years ago we are just approaching the limit of C^{14} dating, and the events between this date and the onset of global deglaciation about 15,000 years ago are capable of being resolved on land as well as in the deep-sea record. 75,000 years ago the Laurentide Icesheet expanded 1,000km or so to the south and west and merged with the Cordilleran Icesheet, which was located over the mountains of western Canada and Alaska. In Europe the Scandinavian Icesheet thickened and grew over the mountains of Norway and Sweden. It is not known if the UK was glaciated at this time or not; but the evidence favours the negative.

The world's climate between the glacial stadial 75,000 years ago and the slightly larger one 18,000 years ago is difficult to understand. The deep-sea work would indicate only a slight amelioration of the ocean temperatures. In North America, the Cordilleran Icesheet retreated and disappeared, and the Laurentide Icesheet retreated north and east to its stable position, although ice still blocked the Saint Lawrence River. In Europe the evidence for this time is not definitive, but finds of buried soils and mastodons in Norway and Sweden suggest that the great mass of the Scandinavian Icesheet had melted back at least into the high mountains.

These interstadial conditions lasted from 60,000 to 25,000 years ago. Botanical and zoological evidence from the southern margins of the icesheets suggests that in places the climate was similar to that of the present; whereas at other sites the records indicate a cooler climate. This long interstadial may have played an important rôle in the final evolution of *Homo sapiens* and must certainly have influenced the migratory movements of our ancestors. During this period, world sea-level may have been about 50m lower than at present. Large areas of continental shelf would thus have been exposed and may have presented favorable habitats for Man and big game animals. Recent work from the Canadian Yukon confirms that early Man wandered into North America during this interstadial—this is at least 19,000 years before the date that many archaeologists accept but the odds are now in favour of a peopling of North America during the major interstadial of the Wisconsin Glaciation. This evidence also supports a series of radiocarbon dates that varied between 19,000 and 26,000 years ago for archaeological sites in Central and South America, something which had caused problems in archaeological circles because the dates found preceded the unequivocable evidence of a rapid peopling of North America about 11,000 years ago by big game hunters.

The last major glacial event of the Last Glaciation was the rapid growth and expansion of icesheets over Scandinavia, the UK, and the Cordilleran mountains of North America, as well as on the southward and westward margins of the Laurentide Icesheet. From available data it appears that this last major glaciation commenced about 25,000 years ago and, on a global scale, the maximum peak of glaciation was reached by 18,000 years ago. However, on the local and regional scales different margins of different icesheets reached their glacial maxima at times that varied between 22,000 and 9,000 years ago: in a general sense the southern margins of the northern hemisphere icesheets reached their maximum close to 18,000 years ago, whereas in the high northern latitudes researchers have argued that the maximum was not attained until 10,500 years ago in Spitsbergen; 10,000-11,000 years ago in East Greenland; 8,000 to 10,000 years ago in the Eastern Canadian Arctic; 13,000 years ago in the western Canadian Arctic; and 14,000 years ago on the southwest margin of the Cordilleran Icesheet.

The World 18,000 Years Ago

Scientists from around the world have been involved over the last five years with the reconstruction of the world's icesheets, ocean, atmosphere, and surface vegetation at the time of the last glacial maximum of 18,000 years ago. A considerable part of this effort has been directed by a group of researchers working under the CLIMAP Project. Funds for the research have primarily come from the USA under the umbrella of the International Decade of Ocean Exploration (IDOE).

The nature of the 18,000-years-ago world, and its aftermath, created or contributed to many aspects of the world that we live in today. The task of reconstructing the world 18,000 years ago from the evidence present today has several different elements. Many of these can be viewed as the synthesis of existing data, but others required the development of new approaches and the use of sophisticated computer simulations. Because it is this type of joint effort which might ultimately lead to a fuller understanding of our present and future climate it is worth spending some time on the approach in order to discuss its strengths and weaknesses.

Icesheet Reconstruction

It might be assumed that icesheet reconstruction is an easy task. However, there are three aspects to such a programme—mapping the margins of the world's icesheets at this one point in time; reconstructing the three-dimensional geometry of the ice masses; and assessing the volume of the world's water that is tied up in the icesheets—and it turns out that none of these tasks is easy. There is not one upon which there is a general scientific consensus.

There are two quite different worlds that people imagine for 18,000 years ago.

In one model—the maximum-ice-extent model—proposed by Terry Hughes and George Denton from the USA and Misha Grosswald from the USSR, a massive icesheet/ice shelf complex completely dominated the geography of the northern hemisphere. They suggest a 3,000m-thick ice dome located over the Barents Sea with additional icesheets on the shallow marine shelf of eastern Siberia. Elsewhere in the northern hemisphere the icesheets are envisaged as flowing seaward across the continental shelves, and there merging with ice shelves in Baffin Bay and the Norwegian/Greenland Seas. In this world the number of biological refugia, or places where plants and animals survived the last glaciation, were extremely limited. Tundra plants and animals could have survived in the relatively small ice-free areas in the Canadian Yukon or parts of

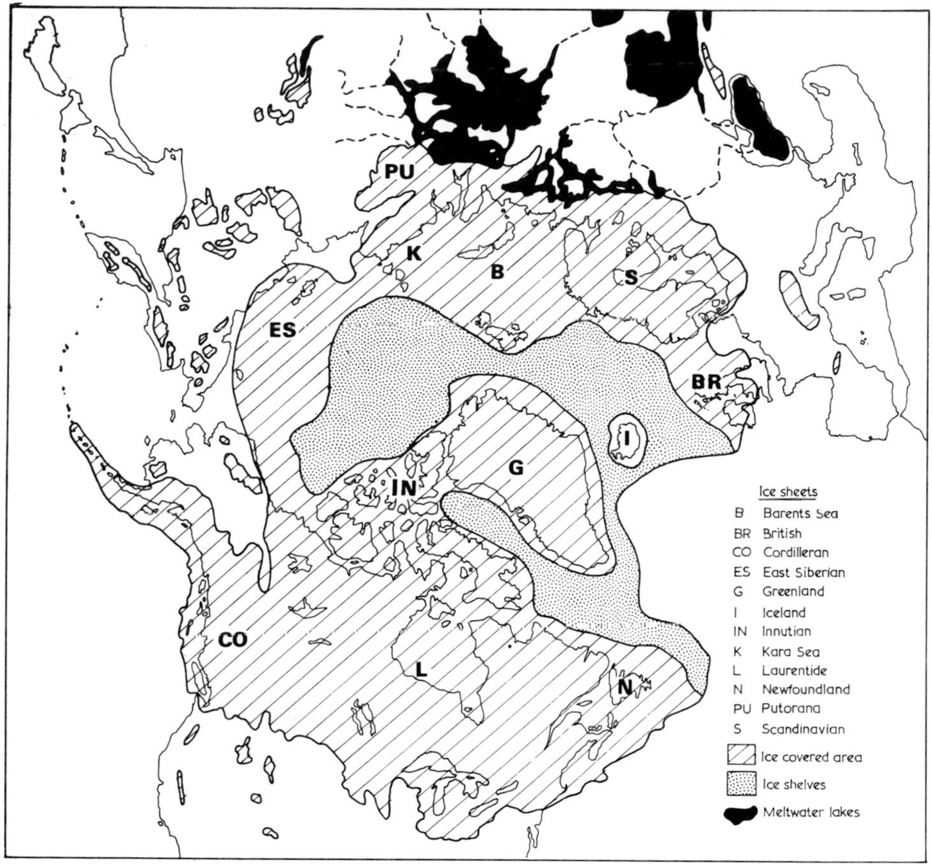

The 'maximum model' for the northern-hemisphere glacial cover of 18,000 years ago.

Alaska, or else of necessity migrated south to inhabit a rather narrow tundra zone which lay south of the Laurentide Ice Sheet. The fossil evidence indicates that in Europe a much broader belt of tundra existed south of the icesheets.

An alternative model stems from research in Spitsbergen, Greenland, and Arctic Canada at sites which on the maximum model lay beneath 1,000m or more of ice. However, rather than finding sediments which are all less than 18,000 years old along these coasts, a large number of different investigators have described a stratigraphic sequence consisting of a marine unit, often rising to 70-80m above present sea-level, in which there are marine shells dated as being more than 40,000 years old; these sediments are in turn overlain at elevations usually less than 20m by another marine stratum which contains marine shells with radiocarbon ages of between 11,000 and 8,000 years. The critical part of all these sections is the hiatus between marine depositions of about 10,000 years ago and of about 40,000 years ago.

The maximum 18,000-years-ago model has to assume that glacial ice extended over these sites but left no deposits. The icesheet is thus considered to have been cold-based (Chapter 3) during the middle Wisconsin Glaciation. The minimum-ice-extent model envisages ice-free enclaves between the main icesheets and either ice shelves or permanent sea ice. In terms of area and ice volume the

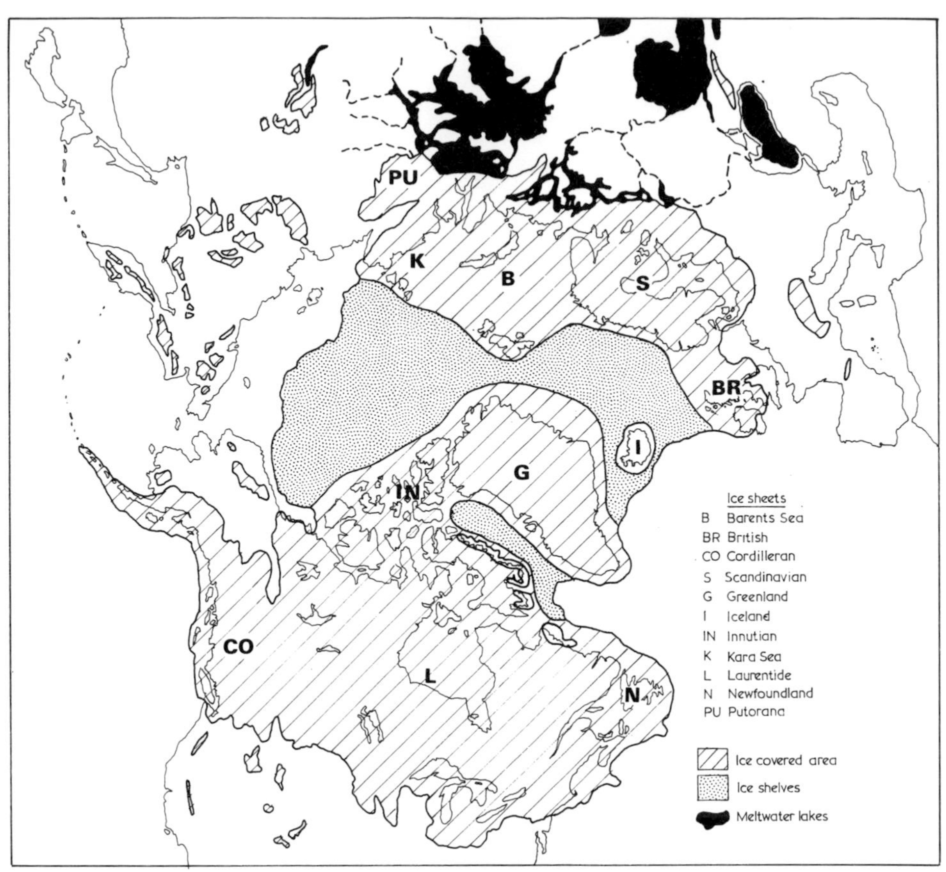

The alternative 'minimum model' for the northern-hemisphere icesheets and ice shelves of 18,000 years ago.

disagreement in North America and Greenland does not matter a great deal. However, the problem of whether or not there was a major icesheet over the Barents Sea and east to the Kara Sea is clearly more significant.

In the reconstruction of these icesheets, workers have used our knowledge of the form and physics of present icesheets to model the ice thickness of 18,000 years ago. However, there are serious problems associated with this exercise. These can be seen when we review the assumptions that have to be made: that the icesheet is in equilibrium; that present icesheets provide analogues for all past icesheets; and that the glacial depression of the Earth's crust has reached equilibrium.

The first and third are probably false assumptions. We now know that an icesheet can build up to its maximum in a few thousand years (say 5,000), and this is too fast for the icesheet itself to reach equilibrium or for the Earth to take up an equilibrium depression.

In addition, Geoffrey Boulton of the University of East Anglia has questioned the validity of the second, most critical point. Icesheets can be reconstructed in terms of their thicknesses by comparison with present-day icesheets which appear to rest largely on rock. However, as Boulton has pointed out, many of the Pleistocene icesheets extended across areas underlain by sediments including impermeable clays and shales and also across wide areas of thick permafrost. Beneath an icesheet spreading across such terrain, the water from any melting at the base could not permeate through the underlying sediments, and hence the frictional contact between the glacier and the bed would be decreased; this is the sort of situation that occurs when a glacier surges or advances catastrophically at a rate of several kilometres per year. With reduced friction at the base (the trapped water acting as a lubricant) the icesheets would be thin, and certainly much thinner than reconstructions based on all three assumptions noted earlier.

There is evidence to support Boulton's ideas in the very low profile of the Laurentide Icesheet in the vicinity of its southwestern margin; and his ideas would fit perfectly the picture of the icesheet in the map on page 193. The stable interior would have a profile comparable with the area of the Greenland Icesheet today. However, the unstable part of the icesheet is bounded at its inner margin by the contact between the permeable granites of the Canadian Shield and the more impermeable Cretaceous sediments further south. In a similar way, Boulton speculates that the ice lobes which penetrated southward in the Irish and North Sea basins may also have been thin.

It is important to note that all the above assumptions tend to ensure that the estimated volume of ice 18,000 years ago is a *maximum* estimate. This volume converted to water equivalent and divided by the area of the world's oceans should give a figure for the world-wide fall of sea-level at the time of glaciation. As we shall see shortly, this simple idea becomes complex in reality. Current estimates vary by a factor of two, the minimum being about 80m and the maximum about 160m.

In the reconstruction of the world's climate during the glacial maximum of about 18,000 years ago the other important inputs into the computer models are the extent of the different vegetation belts, and, more importantly, the sea-surface temperature. One of the critical breakthroughs of the last decade has been the establishment of techniques that enable us to derive, from the changes of the fossil populations of foraminifera in different strata, estimates of the surface temperatures and salinity of the oceans at the times those strata were laid down. The level in the deep-sea cores corresponding to a time of 18,000 years ago is determined by oxygen-isotope stratigraphy and the sea-surface temperature is derived from equations which link specific present-day assemblages of foraminifera with their appropriate ocean environments. We assume that such relationships held true in the past, as related to us *via* the medium of the deep-sea core samples. The results of these studies indicate that during the last glacial maximum sea-surface temperatures were about the same as they are today in the subtropics and tropical oceans but that significantly lower temperatures occurred in the mid-Atlantic between 60° and 40°N; the temperature gradient between the northern latitudes and the equator was thus greater than it is today.

The extent of ice, the changes in vegetation and the sea-surface temperatures; these are the essential pieces of information needed in the construction of large computer models of the general circulation (called GCMs for short). Several models are available and, since each is programmed differently, there are variations between the results obtained from them. However, most impressive are their broad similarities. Our world 18,000 years ago was universally drier.

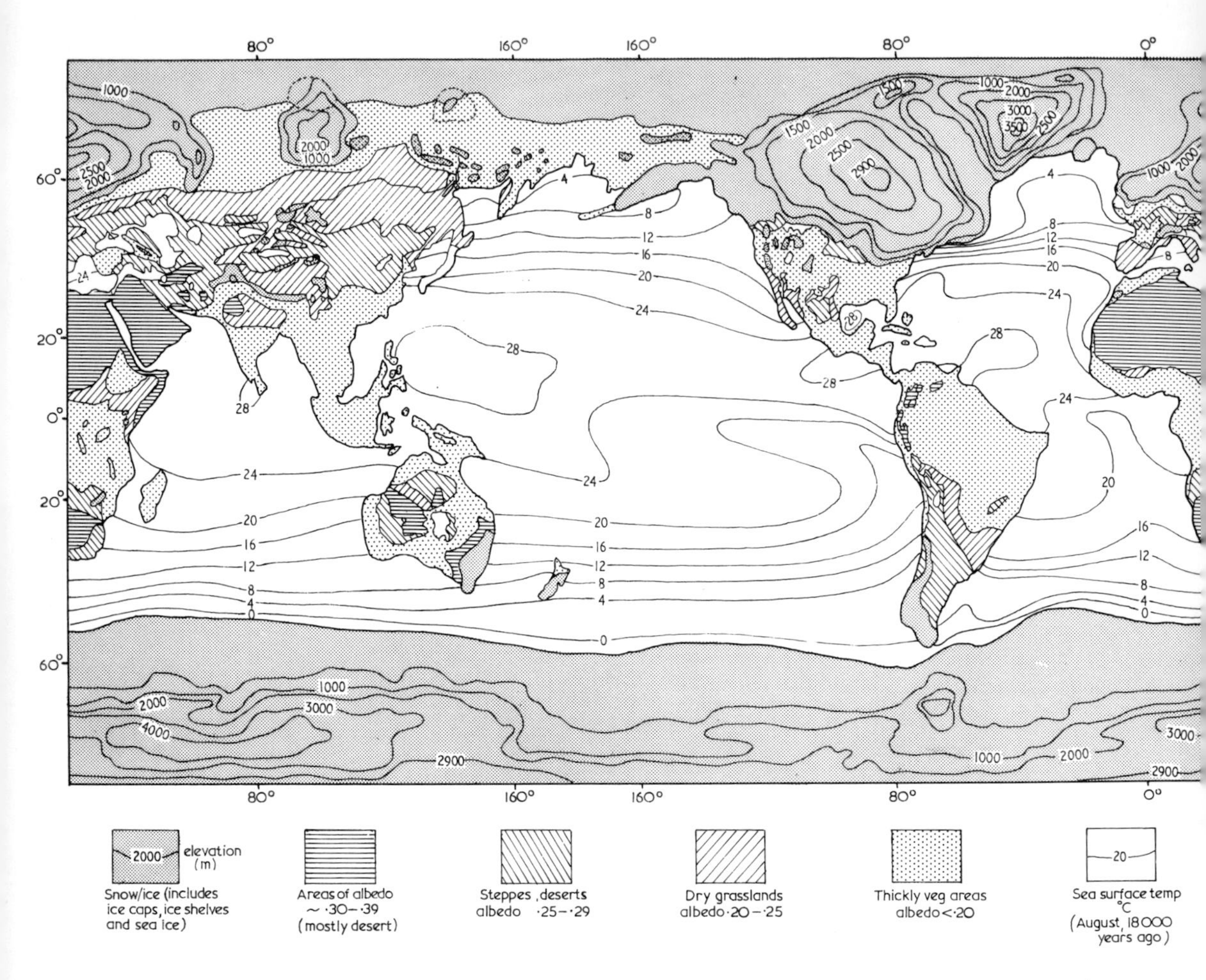

The world 18,000 years ago. The CLIMAP reconstruction of the global ice cover, vegetation types and sea surface temperatures.

Temperatures were much colder over the ice caps and close to the margins of the icesheets, especially in the USSR, but there appears to have been little change nearer to the equator. The GCMs show an atmospheric circulation in equilibrium with the factors mentioned above.

The models give an important although limited perspective on the more fundamental question of: why and how did the last glaciation terminate? We have already seen from the oxygen-isotope record that each glaciation comes to an abrupt end shortly before each interglacial. Broecker and Van Donk have called these events, appropriately, Terminations.

Termination I, on the global scale, is dated at about 13,000 years ago. A simple explanation might be that, just after 18,000 years ago, the equilibrium climate as simulated by the GCMs broke down and rising temperatures melted the icesheets, but in fact this was not the case. Evidences for rising temperatures on land are few until after 14,000 years ago, by which time a considerable amount of water had been returned to the oceans. A possible answer stems from the realization in the late 1960s that the wastage of the interior of the Laurentide

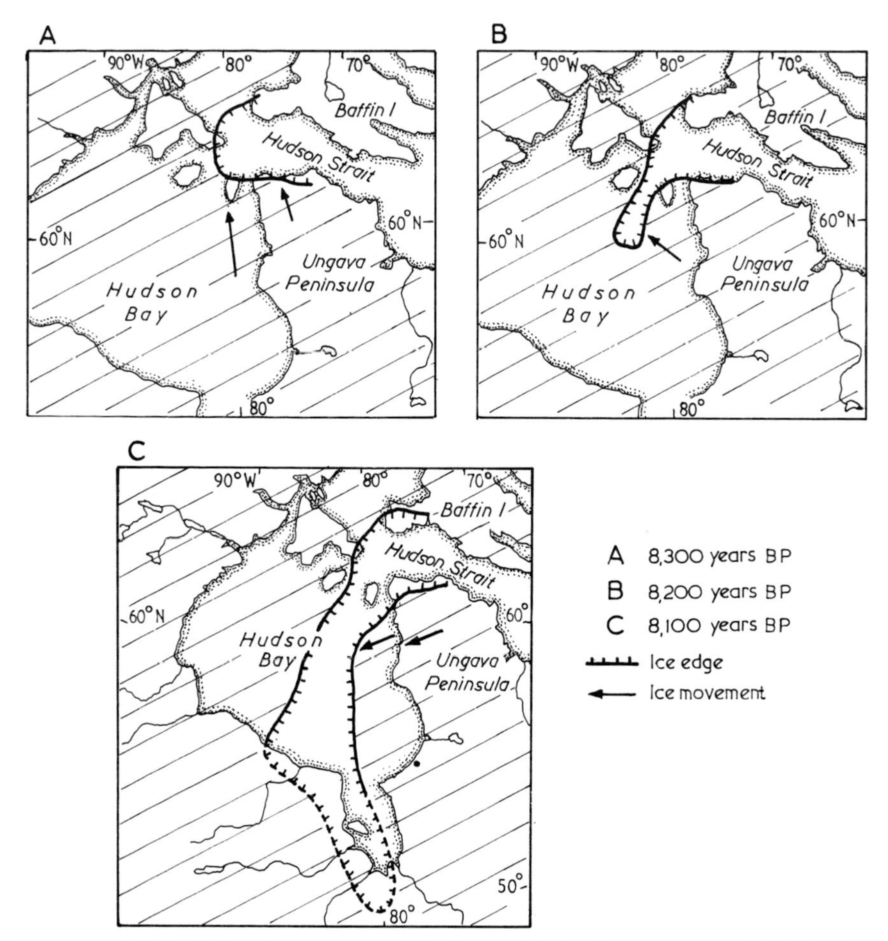

The pattern of deglaciation in the centre of the Laurentide icesheet over a period of 200 years. The rapid breakdown of the old icesheet was caused by the penetration of the sea from Hudson Strait into Hudson Bay, causing the icesheet to split into two smaller remnants.

Icesheet occurred dramatically over a few decades, centuries at most. Such a rapid deglaciation cannot be explained by climatic means but has to involve the mechanical disintegration of the icesheet. In other words, as suggested in Chapter 2, *glaciological* factors assume great importance.

These observations on a former icesheet bear on the problem of the present state of the West Antarctic Icesheet, which is claimed to be close to a state of disintegration. The Laurentide and West Antarctic icesheets have at least one feature in common—the central regions of the icesheets are below sea-level. Hughes, Denton and Grosswald have proposed that such icesheets are inherently unstable. The Scandinavian and Barents Sea Icesheets (if the latter existed) similarly had large areas below sea-level. However, we must bear in mind that the Laurentide Icesheet's stable interior resisted the encroachment of the sea through the Hudson Strait and the Saint Lawrence Valley for tens of thousands of years during the long middle Wisconsin interstadial.

It has recently been suggested that deglaciation is triggered by a combination of isostatic and world sea-level events which affect the margins of icesheets where they end in the sea. If relative sea-level rises at the margin of such an icesheet (either by continued crustal depression or by a sudden rise of world sea-level) then buoyancy would tend to lift the ice front. This would cause calving of icebergs and the development of a calving bay. The retreat of ice would tend to channel ice flow towards the calving bay and this 'positive feedback' might cause the rapid deglaciation that is documented for Hudson Bay.

On the world-wide scale we might envisage the progressive collapse of icesheets, each acting as a trigger to the disintegration of the next. This cascade might be: Barents Icesheet, Scandinavian Icesheet, Laurentide Icesheet, and finally West Antarctic Icesheet. As the icesheets collapse atmospheric circulations should shift and some icesheets, such as the Corderillan, would melt because of the rise in global temperature. Although this explanation might sound complex it is possible that the cause(s) of the Terminations are indeed just that—highly complex!

The Present Interglacial and Current Glaciers

By convention, the last major interval of geological time is called the Holocene. It is an important period both for Man and for our present flora and fauna (see later). The boundary of the Pleistocene and Holocene is placed at 10,000 years ago. By and large, the Holocene covers the period of time during which our present interglacial world has been in existence.

However, we should not imagine that everywhere simultaneously became 'interglacial' 10,000 years ago. In North America at that time the Laurentide Ice Sheet still covered an area of 7.5 million square kilometres, and the last remnants of the icesheets over Labrador and Keewatin did not disappear until 6,000 years ago. In Scandinavia, the ice was just withdrawing from a major readvance 10,000 years ago, but by 8,500 years ago had disappeared from all but the highest cirque basins.

Studies indicate that the Holocene can be subdivided into three broad climatic phases. The boundaries of these vary around the world by a few hundred to a few thousand years, but here we will produce a broad general picture.

The early part of the Holocene (10,000 to 8,000 years ago) was marked by a rapid improvement in the conditions on land and in the oceans. This interval saw the final 'switch' from the 18,000-years-ago world to one which broadly resembled the one we inhabit. In places, possibly many places, the climate may have been warmer and wetter than our present one.

The final collapse of the Laurentide Ice Sheet 8,000 years ago appears to have ushered in a period called the 'Climatic Optimum'. Nearly all records from the northern hemisphere and some from the southern hemisphere indicate that, between 8,000 and 5,000 years ago, average air temperatures were warmer than today's, and, at least at times, conditions were wetter in the deserts of Africa.

In many geological records a major change can be seen starting some time between 5,000 and 3,500 years ago. This transitional period may reflect the gradual lowering of global conditions towards a threshold which has led to 'Neoglaciation' (that is, new glaciation) in many polar and alpine regions. Calcer has suggested that the renewed glaciation of the last 5,000 years has in effect spelt the end of the Holocene Interglacial, and that we are already witnessing what Nils-Axel Mörner of Sweden has dubbed 'The First Future Glaciation' (see Chapter 8).

People have attempted to find a global pattern to the glacial advances of the last 5,000 years. Denton and Karlen suggest that there is a fundamental 2,500-year periodicity to glaciation during the Holocene. The work of my colleagues in Baffin Island, however, suggests a shorter periodicity, and this is also shown by work in the Swiss and Austrian Alps.

The world's present glaciers and icesheets represent responses to climatic changes of considerably different wavelengths. We have already seen that the Antarctic and Greenland icesheets are stable over time periods of a million years (excepting West Antarctica). In the Canadian Arctic the Barnes, Penny and Devon Ice Caps, plus several ice caps in the higher arctic islands, contain ice that dates from the last glacial stage. By contrast, many of the glaciers in the Rocky Mountains, the Alps and elsewhere may date from the last major glacial episode, the so-called 'Little Ice Age'.

Glacier Régimes and Glacial Processes

It is important to remember that there is no single glacier régime. As noted in Chapter 3, individual glaciers respond to changes in their mass balance. Also, no two glaciers have the same turnover, and there are great differences in their activity index (or potential for erosion). The relationship between 'glacier power' and rates of glacial erosion has already been discussed in Chapter 3, but when considering the present ice age we should remember that any individual glacier will have experienced variations in its capacity for erosion during the course of the late Cenozoic glaciations. For example, the massive cirque basins of Antarctica have been ascribed to a period of intense glacial erosion during the middle Tertiary, when the continent of Antarctica was becoming progressively glacierized. Thus, at that time, the climate of the area was not as severe and the capacity for erosion of the initial glaciers was probably comparable to those of the coastal mid-latitudes today. Thus the amount of work (in the form of erosion) done over a period of time by these Antarctic glaciers might look like curve A on the diagram below. By contrast, a cirque on Baffin Island in the Canadian Arctic might produce a curve like curve B, and this can be compared with the rate of work performed in a cirque in Scotland (curve C). These graphs are partly pictorial but carry the important message that any erosional landform is the result of a complex of events changing through time.

Erosion rates of glaciers have not been measured very successfully, although some estimates have already been presented in Chapter 3. If we are interested in the glacial and interglacial oscillations of the Cenozoic we need to know something about rates of erosion and about the variations of these over time. Volcanic rocks in Antarctica have been eroded by cirques at an average rate of about 300mm per 1,000 years. In Washington State, USA, studies of Mount Rainier indicate an overall rate of erosion of 1100mm per 1,000 years. Two comparable studies have been conducted on erosion rates of cirque glaciers in the Colorado Front Range (40°N) and in Baffin Island (66°N). Moraine volumes suggest that in the Colorado Front Range glacial erosion is of the order of 2,000mm per 1,000 years whereas in Baffin Island the estimates are only 50mm

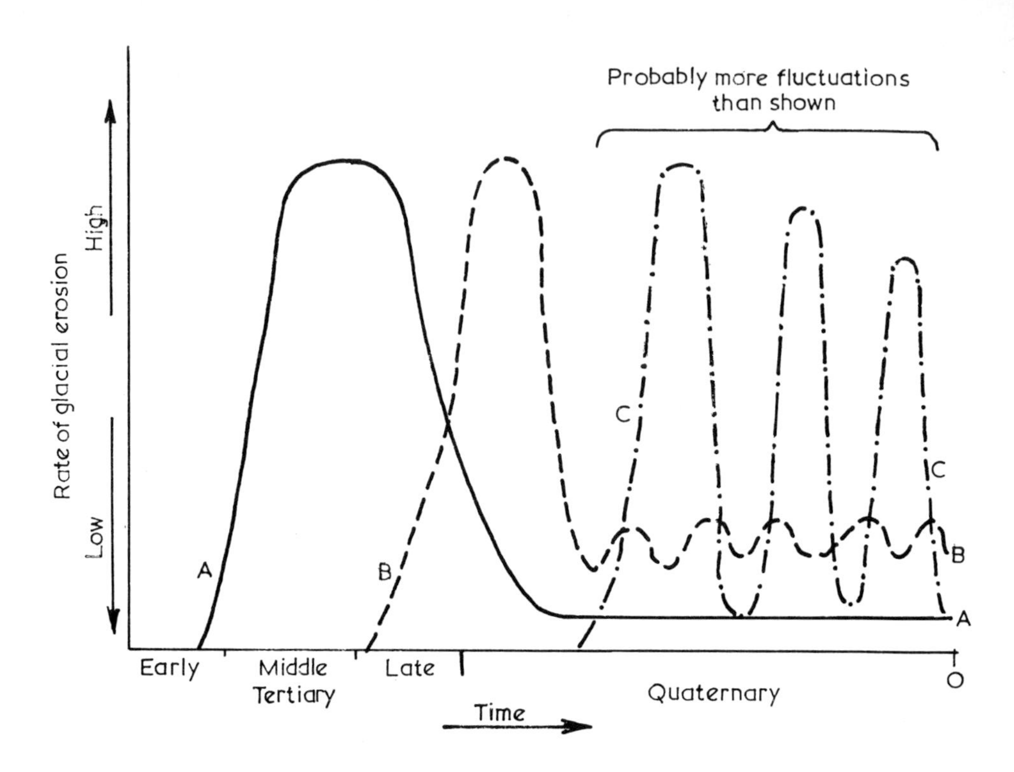

Rates of glacial erosion over time during the Cenozoic ice age. The curves might apply to Antarctica (A), Baffin Island (B) and the Rocky Mountains (C).

per 1,000 years. Thus, in this latter case, the present size of the cirques may indicate erosion continuing over 1 to 3 million years, whereas in the Colorado Front Range the present cirque basins *could* have been cut in about 150,000 years if the measured rates of erosion are applied.

The whole concept of the rate of glacial erosion and its variation through time, on the short (centuries), medium (millennia), and long (millions of years) timescales, is immensely important if we are to understand the way in which the landscapes of our present glaciated world evolved. How much glacial erosion, in terms of metres of rock removed, occurred within the Norwegian fjords 100,000 to 10,000 years ago compared to 1,000,000 to 900,000 years ago? Are we to conceive that glacial erosion considered over the long timescale is constant, or could it be that the majority of our landscapes evolved during the first one or two glaciations and that subsequently little change or modification has occurred?

This is a difficult question to answer. Some modern studies suggest that the Wisconsin or Weichselian Glaciation inherited a world which had already been shaped by glaciation and that little in the way of serious modification occurred. The corollary is that the fundamental character of the glaciated landscape was imprinted early in the Cenozoic Ice Age. It is easy to make such categorical statements—more difficult to either disprove or prove them. In the Canadian

(*Opposite*) A relic of a previous glacial stage. Hare Fjord on Ellesmere Island is an ancient feature, far too large to have been cut by the small glaciers descending from the ice caps and snowfields on either side. In this photograph the fjord waters are covered by sea ice.

Vertical air photo of a cold-based glacier. This is the Lower Gilman Glacier in northern Ellesmere Island. Probably glaciers such as this have only a small impact upon the land surface underneath.

Arctic, evidence to support the statement comes in the form of glacial and marine deposits with ages ranging from a few thousand to in excess of 350,000 years plastered against valley sides at the ends of fjords. Obviously the cutting of the fjords preceded the deposition of the sediments. In Hudson Bay, on Baffin, Banks and Ellesmere Islands, sediments of the last interglacial outcrop in river cuts or just on the present surface. All these areas are believed to have been covered during the last glaciation by 1 to 3km of ice.

This topic raises the ghost of a long-lost controversy—that between the glacial 'erosionists' and the glacial 'protectionists', between those researchers who look at the tremendous scale of a Greenland fjord and ascribe it all to glacial erosion, and those who see the passage of ice over a site as being primarily passive. Like many arguments in the natural sciences both extreme views are on average wrong, but may be correct for specific rare examples. The simple explanation is that at some times and in some places glaciers and icesheets are efficient agents of erosion, whereas at other times and other places they have little or no erosional impact.

Some of these differences in the effect of ice on the landscape are related to the nature of the basal temperature of the ice mass. If the ice is a thin ice carapace covering a summit in the High Canadian Arctic where the mean annual temperature is −18°C it is probable that the ice is cold-based (see Chapter 3), moves very slowly, and serves to protect the underlying stratum. We can contrast such an ice mass with a cirque or valley glacier in a maritime temperate environment. Such a glacier is moving quickly, it is slipping over its bed, and it

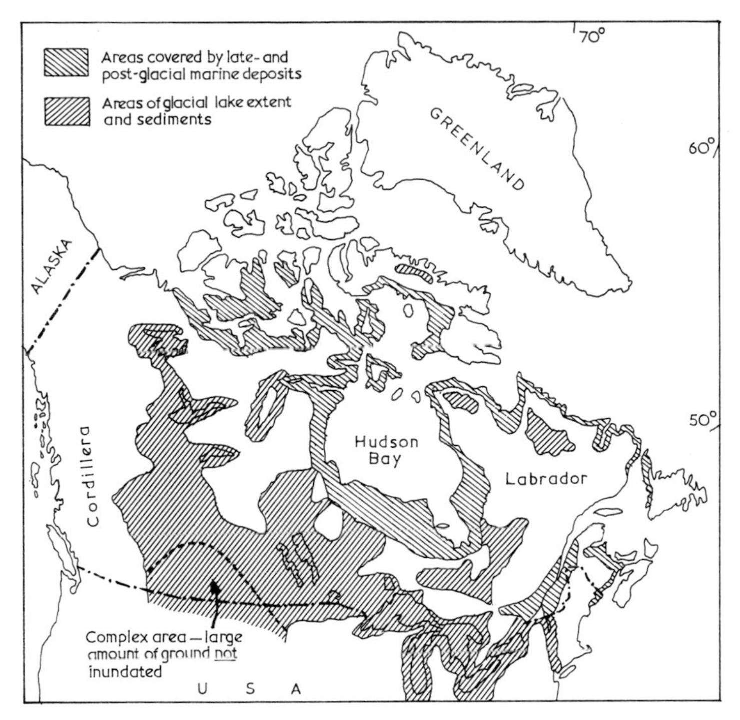

Map showing the areas covered by the sea and by glacial lakes at the end of the last glaciation by the Laurentide icesheet.

is grinding and crushing rock protuberances. It is the landscapes produced by this glacier that we see in the travel posters; such spectacular landscapes characterize the majority of the world's mountain ranges. However, the areas covered by such ice masses in the Pleistocene are small compared to the areas of flat, undulating terrain which once underlay considerable proportions of the Laurentide and Scandinavian icesheets.

Although glacial geologists have wondered and speculated about the grandeurs of glaciated mountains and coasts for about a century, the amount of study and theory to explain glacial landscapes are not as voluminous as those concerned with the nature of glacial landforms. Concerning depositional landforms, advances in understanding have been hindered by the tendency towards specialization. We have become authorities on a limited number of techniques that we apply to all sediments, or else we concentrate our studies on individual landforms such as drumlins or eskers.

A synthesizing concept, proposed by Geoffrey Boulton, deals with the *assemblages* of landforms that would be expected in various situations. Characteristic sets of landforms can be defined for the following glaciological conditions: a valley glacier; a subpolar icesheet or glacier, or possibly a temperate glacier which becomes heavily covered by debris; a temperate icesheet; and a calving margin with the ice ending in either the sea or a glacial lake. The landform assemblages mainly reflect processes that are active at and near the margins of icesheets and glaciers. Geographically the distributions of major areas of glacial deposits are quite restricted in North America and Northern Europe. Viewed as a whole, the area of the former Laurentide Ice Sheet is dominated by vast areas of the Precambrian shield which have been scraped and scoured by the ice. Areas of thick glacial deposits are really restricted to a belt about 300km wide which extends inward from the ice limits.

The major landform assemblages have been outlined in Chapter 3. It is thus interesting to see how important they are in different regions of glaciation. In looking at a glacial map of North America the most striking aspect is the tremendous area that was once covered by water, either because of higher sea-levels during deglaciation, or where the ice retreated into large glacial lakes. This is especially true in North America but also applies, albeit to a lesser degree, in Northern Europe.

Many of these areas which were formerly water-covered are crossed by De Geer moraines (see page 84). These may be the most common glacial landforms in North America. They cover thousands of square kilometres on either side of Hudson Bay, along the arctic coastal plain of Canada, and they have also been described and reported from the glacial lake basins of southern Canada and from the Saint Lawrence Valley.

The second major landform assemblage in North America is the subpolar landform assemblage. Whether the ice was truly subpolar or whether other factors were important is difficult to say. However, it is true that, from the Dakotas up to the Canadian Prairies of Alberta, Saskatchewan and Manitoba, most landforms reflect the 'recycling' of glacial debris through a process of: first the initial accumulation of debris on the surface of the icesheet; then the melting and collapse of the ice through ablation; and finally the deposition of this debris by mass flows. The resulting topography is marked by mounds and hummocks stretching over great distances. Excavation through them for road construction reveals a complex of sediments: glacial tills, lake sediments, and outwash sands and gravels. The subpolar glaciers of Spitsbergen represent an analogue for this particular margin of the Laurentide Ice Sheet.

The third major landform assemblage is usually considered to be 'typical' of glaciated areas. It is not. It is, however, represented in many areas where glacial geology and geomorphology have been under study for a long time! The Great Lakes region of North America is one such area. Here researchers have documented over the last 100 years a succession of glacial deposits which are frequently separated by nonglacial deposits containing evidence for *in situ* preservation. This is, if you like, 'layer-cake' stratigraphy. Not that every former interglacial and interstadial landsurface is preserved, but there are enough well preserved buried soils, peats and forests to be sure that in these areas the glaciers have overridden the surface and deposited additional sediment. The surface of such an area is marked by features which show that the base of the icesheet was at the pressure melting point. Linear features in the form of drumlins and flutings are most diagnostic, and these fan out toward the massive end moraines which mark periods of equilibrium in the mass balance of the icesheet. In places, the meltwater produced at the base of the icesheet has created the long sinuous esker ridges that snake across parts of Canada and the USA.

In the mountains of the Alps, North America, the Himalayas and elsewhere the fourth landform assemblage dominates. Basically the processes and sediments resulting from glaciation are similar here to those occurring at the base of

Vertical air photo of a large valley glacier in the mountains of Greenland. Notice the steep gullied rock walls on either side of the glacier and the very prominent moraines on the ice surface. This glacier is unusual in that it splits into two in its lower part.

a temperate icesheet. However, the restriction of the glacier to a valley and the presence of large areas of steep rock walls in the upper basins provide distinctive elements; in particular, the rock-falls form a contribution to an accumulation of debris on the surface of the ice, a contribution that is not available to icesheets; much of this material is incorporated into lateral, medial and end moraines. Valley glaciers in many areas of the world are wet-based (Chapter 3). They leave behind them lineated surfaces in either drift or bedrock. In places such as the Rocky Mountains of the USA and in North Italy the areas of the former late-Pleistocene glaciation are bounded by massive lateral and end moraines which tower 60 to 100m above the valley floor. Upvalley, much of the volume of material in the moraines is derived from direct glacial erosion of the bed.

Icesheets and Sea Levels

One of the most important effects of a glaciation is in terms of local and global sea-level changes, as already described in Chapter 3. With a view to understanding the Cenozoic interrelationships of glacio-eustasy, glacio-isostasy and hydro-isostasy, a group from the University of Colorado has been using a computer to simulate what happens to sea-level at any point on the Earth's surface. As a basis they have used the known history of the world's icesheets and the major characteristics of Earth structure. In addition, they have found it necessary to include in their computations the gravitational effect on sea-level of the mass itself. (This was first proposed by Woodward in the late 19th century. The mass of the icesheets is so large that they literally attract the surface of the oceans so that they 'climb' toward the ice margin by as much as some 30m or so.)

The computer results have shown that the effect of the deglaciation from the last glacial maximum 18,000 years ago was to produce different sea-level 'signatures' around the world. The research group identified five basic patterns of sea-level movement on a global scale. In areas inside the perimeters of the former icesheets the dominant process over the last 18,000 years has been one of uplift, as the Earth's surface has risen toward its preglacial altitude. During a glaciation the displaced material from under an icesheet gives rise to a bulge which encircles the icesheet margin; as the icesheet decays, the bulge region sinks. Further away from the ice loads, namely in the ocean basins, the story is one of continued rise of sea-level and submergence of the coast as water is returned to the oceans. However, in the southern hemisphere the simulation suggests a period of land emergence of 1 to 2m commencing about 5,000 years ago.

The insights into local sea-level data afforded by the computer modelling are considerable. However, it is worth noting that the model is still not totally realistic. For example the effects of Antarctic Icesheet history have yet to be included, and in addition the current model assumes the Earth does not have a lithosphere. Nevertheless, the message from this work is plain—sea-level at any point on Earth is *not* simply a reflection of the amount of water added or subtracted by the unstable icesheets.

In a general way we can compute that, during the maximum of a glaciation, sea-level away from the icesheets is lowered by 80 to 150m. This is important because it indicates that during glacial periods large areas of the world's shallow continental shelves were dry land and hence could be inhabited by plants and animals, including Man and his early ancestors. Anywhere from 1 to 400km of extra terrain was added to the world's land surface at the coasts, and this compensated in part for the land lost under the icesheets. Proof of the use of these areas by plants and animals is readily available through coring and explorations on the continental shelf. Remains of plants, nearshore shells, and parts of large mammals such as mastodons are frequently dredged up by inshore fishing vessels—as well as research vessels, of course.

Late Cenozoic Glaciations, Plants, Animals and Man

Although we know that the flora and fauna of the world have changed dramatically over the course of the late Cenozoic glaciations, it is not at all easy to obtain a quantitative estimate of the degree of responsibility of the glaciations observed for these changes.

As noted in Chapter 1, we can infer that during more favourable climatic periods, when environmental stresses were less, evolution may have progressed more slowly and actual extinctions of large numbers of plants and animals were probably rare. This is partly because, during stable conditions, ecological niches are soon filled and thus the opportunities for the development of new species are restricted by competition from existing ones.

Our knowledge of past lifeforms is always limited by their preservation and accessibility in the fossil record. However, it is estimated that only 10% of the present plant species of northwestern Europe date from periods prior to the

Miocene, and R. F. Flint notes that of 119 mammal species now existing in Europe and nearby Asia 113 date from the Quaternary and of these 6 evolved as recently as the last glaciation—about 100,000 years ago.

The Cenozoic history of the world has really been dominated by two significant events that would effect evolutionary trends. The first is that of a lowering of temperatures, and the second the break-up of landmasses and the development of large ocean basins. The latter process has resulted in the compartmentalizing of the faunas and floras on the continents and islands. Some movement of faunas between Asia and the Americas was possible at times during the late Cenozoic glaciations across the shallow Bering Shelf, but this is quite far north, and thus the influx of animals from Asia to the Americas, and *vice versa*, would have been limited to northern boreal and arctic forms. There is only limited evidence that the lowering of temperatures and the repeated waxing and waning of icesheets resulted in specific physiological adaptations. It is hypothesized that the animals that lived in northern latitudes developed blubber and wool, whereas the development of deserts caused food and water storage adaptations in camels and related groups.

What problems would the repeated Cenozoic glaciations of North America, Europe and elsewhere pose to plants and animals?

This topic has already been considered in a general way in Chapter 3, but here we can make some more specific points concerning the Quaternary glaciations. With each expansion of the ice the major impact would be on the northern species. Not only would temperatures become lower during a glaciation and moisture less freely available, but the actual space occupied by these species would be severely reduced because large parts of high- and mid-latitude areas would be covered by permanent ice. In order to survive, plants and animals would have to migrate to refugia where they could 'wait out' the glaciation.

We can speculate about several different kinds of refugia that were exploited during late Cenozoic glaciations. First, we can postulate that during glaciation the vegetation belts of the northern hemisphere simply shifted southward. Thus, in North America, the southern boundary of the tundra migrated from approximately 60°N to lie south of the icesheet at 40°N. However, research has indicated that in North America during the glaciations the tundra zone was restricted and may have been only 50 to 200km wide—compared to 1,500km today! Thus the concept of migration in the simple sense suggests a crowding and mixing of the flora and fauna. In addition, the translation of tundra to 40°N would result in physiological differences due to such factors as Sun angle and day length.

Fossils and the distribution of modern plants, beetles, and animals have been used to argue for the existence of refugia at high northern latitudes. The best documented example is the large unglaciated area of the eastern USSR, Alaska and northwest Canada which is termed Beringia. Beringia consisted during glacial periods of the lowlands of the Bering Shelf and the interior lowlands and uplands of central Alaska and the Yukon Territory. Even the mountains of this large area were not totally ice-covered. Beringia thus offered a wide variety of habitats that could be used and exploited by plants and animals displaced by the growth of the large icesheets. Other northern refugia are not as well documented, and as a result there are considerable arguments as to whether there

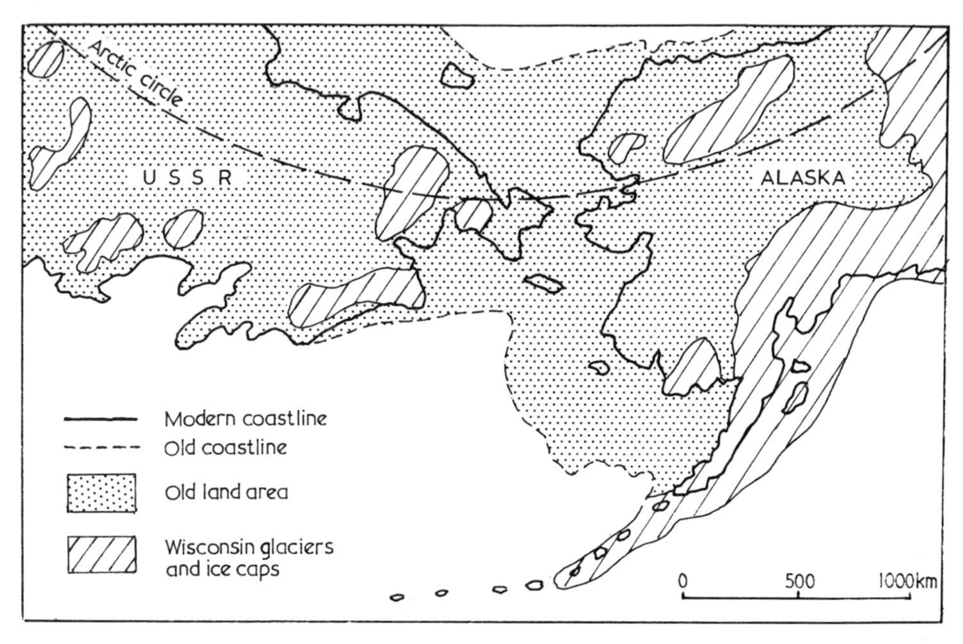

Map of Beringia, the most carefully studied of the areas believed to have been plant and animal refuges on the peripheries of the Arctic Ocean.

were any during the late Cenozoic glaciations in Arctic Canada, Greenland, Iceland, Scandinavia and the arctic islands fringing the Barents and Kara Seas. Botanists and zoologists have argued in favour of them, basing their arguments on the disjunct nature of several key biogeographic distributions of plants and insects; their hypothesis is commonly referred to as the 'nunatak hypothesis'. Many people have been sceptical about the ability of plants and insects to survive winter in these northern latitudes during the height of a glaciation. However, it is worth noting that a Norwegian Botanical Expedition reported over 60 species of vascular plants on some nunataks about 150km in from the Greenland ice margin. The survival of this relatively rich flora might, however, be a local effect, due to the amount of sunlight prevailing on a small piece of rock surrounded by a highly reflective surface (see page 50).

However, the idea that a significant proportion of arctic organisms survived on mountains fringing the icesheets is probably unrealistic. As we discussed earlier there is some evidence, at least for the Last Glaciation, which indicates that large icefree areas existed during the 28,000-year glacial maximum—areas such as Peary Land in North Greenland, parts of Banks Island in Arctic Canada, and the Alaskan North Slope may have been available as refugia. In addition, if we side with the glacial minimists (page 197), there is the strong likelihood that large areas of the continental shelf would have been exposed during times of global glacial maxima: these low plains would have provided additional space upon which some of the organisms might have survived the rigours of a glacial age.

At the moment we are not sure of the relative importance of the three basic kinds of refuge, namely: areas south of the icesheet; areas close to the icesheets at northern latitudes, such as Beringia; and mountains and uplands forming

A hill summit or nunatak completely surrounded by highly crevassed glacier ice near the edge of the Greenland icesheet. Such nunataks may have provided refuge for certain plant species during the Cenozoic glaciations.

nunataks surrounded by glacial ice. Probably the last mentioned are least important, but the present situation in Greenland indicates that they cannot be omitted from any serious analysis of the problem.

The pattern of repeated glaciations followed by the rather shorter interglacials of the late Cenozoic has slowly whittled away at the genetic variability of the Earth's lifeforms.

Palaeontologists have divided the Quaternary into broad faunal units. The oldest of these is the Villafranchian fauna, which can be recognized as starting in the Lower Pleistocene and possibly even in the Pliocene. The fauna comprised a group of animals that were often small in stature compared with their later relatives. True elephants, horses and cattle were new elements which typified the Villafranchian.

A major change occurred 900,000 to 600,000 years ago between the Waal and Cromer interglacials. At this time a large number of species became extinct and entirely new elephants and rhinoceroses appeared. Large fossil faunas of this Middle Pleistocene group have been recovered from sites in Germany. The extinctions of many species coincided with the appearance of the first mammals to have an arctic or 'cold' affinity: bones of reindeer, musk-ox, and lemming occurred for the first time in the stratigraphic record, and later evolution produced the woolly mammoth, woolly rhinoceros and moose.

The Middle Pleistocene faunas continued, with modifications, until the Last Interglacial, about 125,000 years ago. The major change was the final extinction in high latitudes of forms such as the hippopotamus, which in our present view seem ill-suited to swimming in the Thames and other such northern rivers! The Upper Pleistocene faunas of northern lands were dominated by the large grazers, such as various types of mammoths, mastodons and bison. In addition, there were a large number of different ungulates occupying the steppe grasslands.

A last major wave of extinction occurred at the late-Pleistocene/Holocene boundary, about 10,000 years ago. This wave of extinction exceeds even the present 'kill' of rare and endangered species that can be directly attributed to one of the last mammals on the scene, *Homo sapiens*. The causes of the widespread extinctions at the close of the Last Glaciation have puzzled many researchers. Why—after surviving throughout a glaciation and in many cases surviving glaciations—did a particular species (or even a whole genus in the case of *Mammuthus* and *Mammut*, the mammoth and mastodon families) not continue to survive and enjoy the more equable climates of the last 10,000 years? There is little to suggest that the present interglacial is different in kind from earlier ones; if anything, it ranks as a cool interglacial, and certainly this ranking is correct when viewed against the Last Interglacial. To a degree each interglacial was unique as regards the migration patterns into and out of the refugia, which were never totally similar. Thus it might be argued that some specific, unique combination of climate and vegetation response led to the extinction of the large browsers which had formed such a major element of the upper Pleistocene fauna.

Paul Martin has been one of the most vocal advocates of a non-'natural' cause for the extinctions. He has argued rather well that it can be no coincidence that the extinctions, particularly in North America, of the northern line of elephants, many species of bison, and several deer-types occur at a time which, according to the archaeological evidence, saw the advent of Man the big game hunter on the North American scene. Archaeological sites in the Great Plains of North America dating from between about 11,000 and 10,000 years ago do show undisputably that these early people had the technology required to hunt and kill large game animals. Skeletons with spear points embedded in them have been unearthed. In addition, kills were carried out by fall-traps and by driving herds of animals over cliffs. By 10,000 years ago Man was living in close-knit, organized communities, probably organized by the need for group planning of hunts. Martin feels that, through selective killing of young animals or female animals, the populations of game animals were quickly reduced to critical levels so that the reproductive capacity of the species fell below the level necessary to maintain it.

It is probable that other factors worked toward the extinction of species around 10,000 years ago, but the coincidence with the arrival of Man with an efficient killing technology is hard to ignore.

The importance of Man and his direct ancestors within the context of late Cenozoic ice ages is not simple to interpret. The Russians have made the most

The carcass of a baby mammoth found in 1977 in the muds of the Kolgine River, Yakutsk Republic, USSR.

A hut constructed from mammoth bones, Mazhirick Village, Ukraine. During the late Stone Age, as the climate gradually improved, organized hunting contributed to the extinction of the Eurasian mammoth herds.

direct geological association and have called the last 2 to 3 million years of Earth history the *Anthropogene*, thus emphasizing the development during the Pliocene and Pleistocene of Man's direct ancestors. But it is very difficult to demonstrate conclusively that the repeated glaciations of the northern hemisphere and the gradual evolution of *Homo sapiens* are respectively cause and effect.

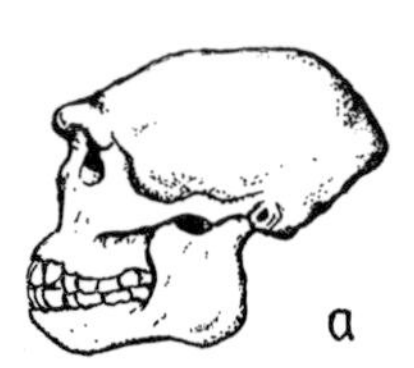

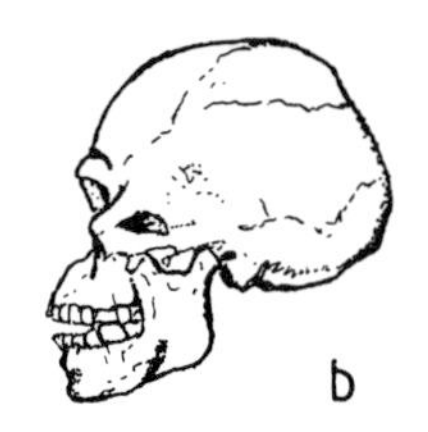

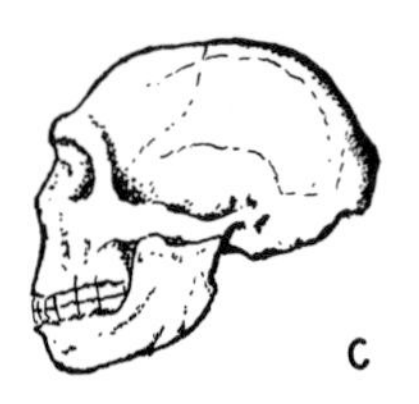

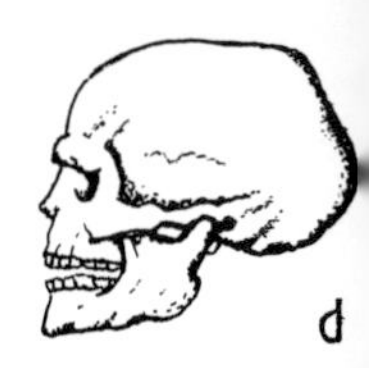

The evolution of Man during the Pleistocene. The skull shapes of (*a*) Java Man, (*b*) Early Neanderthal Man, (*c*) Late Neanderthal Man and (*d*) Cro-magnon Man show a number of stages through which *Homo sapiens* has passed along the way.

The origins of Man have been traced to a small hominid, 1.2m tall, called *Ramapithecus*, found in northern India in sites dated at about 9 million years old. Some 4 to 6 million years later this genus developed into the group of hominids called *Australopithecus*. This group developed along two evolutionary pathways —one became extinct, the other led through *A. habilis* to *Homo erectus* and then *Homo sapiens*.

Australopithecus has been found only within the African continent and it appears from the evidence currently available that the cradle of mankind was on that continent. Glaciations of the Eurasian continent were undoubtedly reflected in environmental changes in Africa, but, as climatic reconstructions for the last 18,000 years have demonstrated, we can no longer make simple equations of glaciations and climate. Thus radiocarbon dating has shown that the periods of higher lake levels in Africa (the so-called pluvials) are not synchronous with glaciations in Northern Europe. In addition, the modelling efforts by scientists using GCMs have strongly suggested that the amount of change in the subtropics and tropics during a glacial maximum is quite small. Possibly the most significant fact that has emerged from these attempts to simulate world climate during an ice age is the overwhelming trend for the predictions to indicate that glacials are equated with periods of reduced global precipitation. At least in the last glacial maximum it is suspected that one of the environmental regions most affected was that of the Tropical Rain Forest, which shrank considerably during the Wisconsin Glacial maximum. These data may be applied cautiously to the likely changes that would have involved *Australopithecus* and *Homo* species during lower and middle Quaternary time.

Homo erectus (Java man) was a 1.5m tall hominid who appears to have spread throughout a good deal of the Old World. Remains of this early ancestor have been found in Indonesia, China and Africa. Sometime during the middle Pleistocene, at a time equated with the classical Elster Glaciation and the succeeding Holstein Interglacial, the *Homo sapiens* line became a distinct offshoot of the older *H. erectus* stock. Swanscombe and Steinheim Man are part of the early *Homo sapiens* line and, significantly, many of the finds of these people come from sites in Northern Europe close to former icesheet limits.

The *Homo sapiens* line was quite variable, with our own kind (*Homo sapiens sapiens*) being but one of a number of types that evolved; the others died off. Probably the best known of these was the famous Neanderthal Man (*Homo sapiens neanderthalensis*) who was successful and widespread during upper Pleistocene time but who appears to have become extinct as a type about 35,000 years ago during the middle part of the Last Glaciation. Why Neanderthal Man became extinct in the middle of the Last Glaciation during a period of interstadial conditions is not yet known. It is equally speculative to wonder what interactions, if any, occurred between these people and the more advanced (?) line of *H. sapiens sapiens* from which we are all descended. It is possible that the extinction might have been caused by Man the warrior.

Rapid changes and developments occurred during the latter part of the upper Quaternary. Fossil finds indicate that Man had spread widely from the original homeland in Africa, but it was not until this late period that he finally migrated to Australia and the Americas, while the occupation of New Zealand and the oceanic islands of the Pacific had to wait for advanced marine technology.

The way in which Man spread to the Americas has occasioned much heated debate. During glaciations, sea-level would have been low, so that Asia and North America would be connected across the Bering Sea. Fossil evidence from Alaska indicates that during glaciations much of the area of Beringia was a refuge for plants and animals. Bones and horns of many large grazers, such as the mammoth, have been found in abundance. Early Man, called Palaeolithic after his artifacts, lived and survived successfully in the steppes and plains of northern Russia during the Middle and Upper Pleistocene, so what held the group back from wandering into North America? Suggestions of an early entry of Man (prior to 11,000 years ago) into the Americas has received as much enthusiastic support as did Wegener's idea of Continental Drift! True, many of the finds and many of the dates are not as well fixed geologically and archaeologically as they might be, but it has become increasingly awkward to explain away *all* the cumulative evidence. Until recently some of the best evidence of settlement came from Central and South America where stratified deposits with human artifacts have provided radiocarbon dates that range in age between 14,000 and 26,000 years old. These dates are well before the accepted peopling of North America by the big game hunters that were associated with beautifully worked spear points of Folsom and Clovis design. There thus was a noticeable gap in time between sites of early Man in the southern Americas and those to the north. This led to speculations that the first people did not cross the land bridge, but used boats or rafts. Certain similarities exist between some of the early coastal sites along the west coast of South America and Japan.

The importance of Beringia in the Quaternary saga of North America has been stressed by a major research effort of the National Museum of Man in Canada. The focus of the multidisciplinary investigations is the evidence for worked bones in sites along the Old Crow River in the Yukon Territory. Radiocarbon dates indicate that the bones are more than 25,000 years old. Initially the bones were not found *in situ*, and hence it might be argued that Man merely used fossil material for his tools. However, subsequent fieldwork has located worked bone within the river bluffs. Wood from the same levels gives ages of at least 35,000 years and the interpretation is that a group of people lived in this part of Beringia during the Middle Wisconsin interstadial (this lasted from approximately 60,000 to 25,000 years ago). Whether they managed to survive the climatic deterioration that led to renewed glaciation in the mountains of the Yukon Territory and Alaska, and the readvance of the Laurentide Icesheet towards the Old Crow Basin, is not known. What is known is that between 11,000 and 10,000 years ago a wave of technically well equipped hunting people

The little church in the deserted settlement of Furufjörður, northwestern Iceland. The climatic deterioration of the Little Ice Age caused many such settlements to be abandoned, and the Icelandic population was reduced by half because of famine and emigration.

occupied the Great Plains of North America from Texas to Alberta. It is very difficult to link these people with a specific invasion of North America across the Bering Land Bridge and so we might safely assume that they were already in residence on the Great Plains before the maximum of the Last Glaciation. This idea is supported by the Alaskan archaeological record as currently known, for it contains no evidence of a large influx of people during the period in question.

The technological and cultural flowering of our species is closely associated with the last 10,000 years of interglacial climate. The association between climate and Man's prehistory and his recorded history has been commented on by many people. Once in vogue, then forgotten, and now hovering on the skyline again is the spectre of Ellesworth Huntingdon's 'climatic determinism' whereby wars, pestilences and civil strife are closely associated with climatic change. In more recent years Professor E. le Roy Ladurie has indicated in his book *Times of Feast, Times of Famine* that Renaissance Man was not immune to these changes, and we know that the Little Ice Age (which has only just ended) made a great impact upon the people of Scandinavia and the European Alps.

The future and present importance of climatic change and the ever present threat of a new glaciation are examined in the next chapter.

The ice reports from the Canadian Arctic for the 1978 summer have just come in. Summer temperatures are 3 to 5 C° cooler than normal; snow is still heavy on the high plateaux of the region; and the ice is still thick and heavy in Baffin Bay. How many 1978 summers does it take to trigger the First Future Glaciation?

JOHN ANDREWS

8

Rhyme, Reason and Prediction

Problems of Interpretation

Throughout this book we have been involved in an exploration of geological time and the ice ages which have punctuated it. Hopefully, the preceding chapters will have given us a more realistic sense of time than we had before. But, as Grant Young pointed in Chapter 4, we have a drastically foreshortened sense of time; we see the Quaternary ice age in great detail, but we have only the briefest of impressions of the Precambrian ice ages. We can speculate about the occurrence of more than one Lower Proterozoic ice age and we may feel that there is good evidence for three Upper Proterozoic ice ages; but before the beginning of Phanerozoic time the evidence is so scanty, and so dependent upon radiometric age determinations which may well be in error, that we can be certain of nothing.

Our problems are compounded when we view the Precambrian as we realize that the old geological 'principle of uniformitarianism' may have little value in the interpretation of the most ancient sediments. The idea that the present is the key to the past, and the corollary that modern environments can be used for the reconstruction of ancient ones, ceases to have meaning when we realize that the Precambrian atmosphere may have had an excess of carbon dioxide and hardly any oxygen.

Stromatolites, red beds and maybe ancient tillites may have to be explained by reference to physical laws which while suspected are as yet not properly defined. Maybe the mechanisms responsible for ice ages were different in Precambrian times; maybe the processes of glaciation were also very different from those described in Chapter 3.

Earlier in this book, when we were examining some of the factors which might be responsible for ice ages, it was extremely difficult to assess the relative merits of a multitude of ice-age theories. We looked at theories involving continental drift and mountain-building, theories involving changing astronomical alignments, theories concerning the behaviour of ice. Some of those theories may have seemed to be mutually reinforcing while others may have appeared mutually exclusive. Some were simple and elegant while others were complicated and so dependent upon 'chance' combinations of circumstances that their universal application seemed dubious to say the least. And yet should we really search for a *universally* applicable theory in order to explain the winters of the world? Should we not be satisfied with the idea that each ice age might have been triggered by its own peculiar combination of environmental circumstances?

Certainly every ice age has occurred at a time of unique continental arrangements, unique oceanic circulations, and perhaps also unique atmospheric conditions. If we look at geological time as linear and irreversible, we have to conclude that no period of the Earth's history has been exactly duplicated by any other period before or since. The preceding chapters have demonstrated forcefully how different planet Earth must have been in Precambrian times, late Ordovician times, Permo-Carboniferous times and during the Quaternary period. But to claim that we do not need to look for an ice-age theory which is applicable to *all* ice ages is to take the easy way out. After all, we have the apparently regular spacing of ice ages, through the later part of geological time at least, to exercise our analytical faculties. At the beginning of this book (page 16) it was argued that ice ages occur at intervals of 150 million years or so. But why are there no widespread signs of a Jurassic ice age where there should be one, about 150 million years ago? Is the periodicity of ice ages more apparent than real? Let us examine this problem in a little more detail now that we have considered the main ice ages.

Periodicity or Chance?

The earliest ice ages, discussed by Grant Young in Chapter 4, are now reasonably well 'fixed' by a large number of radiometric age determinations. They appear to have occurred around 2,200-2,300 million years ago, 900 million years ago, 750 million years ago, and 600 million years ago.

The evidence from the Lower Proterozoic is difficult to interpret, especially with regard to dating. The Huronian glacial deposits are the oldest known; the youngest of the glacial formations in the Lake Huron region is dated as 2,288 million years old, and the oldest one may have been laid down several million years previously. The South African glacial rocks from the Lower Proterozoic are between 1,950 and 2,340 million years old, and the Australian glacial rocks are of virtually identical age. The evidence is suggestive of more than one ice age, but their precise dates are at present unknown.

The 150-million-year interval does seem to apply to the Upper Proterozoic ice ages. On the other hand Grant Young himself, having made a careful study of the evidence from the period 1,200-600 million years ago, considers that the dating evidence is not really strong enough to support the identification of the three distinct ice ages referred to by some others as Gnejsö, Sturtian and Varangian. Hence his cautious approach to the problem of periodicity in Chapter 4. Although the Gnejsö glacial deposits all seem to be about 900 million years old, the later tillites and related sediments are much more varied in age. The Sturtian glaciation seems to have occurred between 810 and 715 million years ago in different parts of the world, while the Varangian ice age is dated to between 680 million years and 570 million years old according to the location of the dated deposit.

Most of the Upper Proterozoic age determinations completed so far have 'standard errors' of between 20 and 50 million years, so all of the above dates are approximations only. Even if we accept them at face value, we have to convince ourselves that individual ice ages could have lasted for as long as 50-100 million years, perhaps with the coldest parts of these ice ages occurring at 150-million-year intervals. Again it is difficult to comprehend the timescale involved. But we should ask ourselves whether ice ages lasting for, say, 50 million years could have 'accommodated' the migration of landmasses from tropical to polar areas or *vice versa*; as suggested in Chapter 4, we have to explain the presence of widespread tillites at all sorts of palaeolatitudes including the equatorial zone.

In answer to this query, the evidence suggests that there is no great problem. We know from the 'recent' end of the geological column that the South Atlantic and North Atlantic Oceans opened within 50 million years, and it took the Indian Subcontinent only a little longer to migrate from high southern latitudes across the equator and into the northern hemisphere tropics. Grant Young has suggested (page 130) that the crust was even more mobile during Precambrian times, and there no longer seems to be any point in worrying unduly about Upper Proterozoic 'tropical icesheets'.

The Phanerozoic ice seem to have been no less persistent. The Ordovician ice age (Chapter 5) seems to have culminated about 450 million years ago, although some of the glacial traces in the Sahara may be several million years older than this. Also, there are tillites of supposed Silurian age in Argentina and Bolivia, and some authorities believe that glaciers may have existed in parts of the ancient supercontinent of Gondwanaland throughout the Silurian period. The timespan of the Late Ordovician ice age may therefore have been something like 50 million years, from about 460-410 million years ago.

The Permo-Carboniferous ice age (Chapter 6) seems to have been so intensive and so prolonged that it has often been subdivided into several distinct ice ages. It seems to have lasted from about 340 to about 240 million years ago—a timespan of 100 million years. If one counts the periods of mountain glaciation during late Devonian and late Permian times—i.e., before and after the main part of the ice age—the total span during which glaciers affected the face of the Earth was something over 150 million years.

Finally the Cenozoic ice age, in which we now find ourselves (Chapter 7), shows every sign of being extremely persistent, since there are large land areas within 30° of the poles in both hemispheres. The beginning of the ice age is usually placed at about 2 million years ago, but we now know that there was a dramatic lowering of temperatures (at least in the high latitudes of the southern hemisphere) about 38 million years ago, during the Oligocene epoch. About 13 million years ago, during the Miocene, the Antarctic Icesheet began to form, and shortly after 4 million years ago it reached its full extent. The Greenland Icesheet first built up about 3½ million years ago. This sequence of events, which appears to be well dated, gives us an analogy for the development of earlier ice ages. It also shows us that the Quaternary ice age is already over 10 million years old and well into its stride.

To return to the matter of periodicity, we can now ask ourselves:

Should we dare to talk about a 150-million-year ice-age 'interval' when we know that some ice ages last for 50 million years or more? There is good evidence that the Varangian and Permo-Carboniferous ice ages each lasted for at least 100 million years, so do we find ourselves in a position of 'overlapping ice ages'? In other words, is it not wiser to talk about glaciation as a normal and persistent feature of planet Earth, with apparent ice ages occurring on continental landmasses only when they happen to wander across the high-latitude zones on their global migrations?

How do we explain the absence of ice ages from certain periods of geological time? Why was there no large glaciation during Jurassic times, 150 million years ago? And why were there no ice ages during the greater part of Lower and Middle Proterozoic time?

These are enormous questions; partial answers have already been given to them in Chapters 1 and 2 and in the chapters dealing with the individual ice ages. It is certainly likely that glaciation in polar latitudes and in high mountain areas all over the world could have occurred during any part of geological time. It is also likely that polar and mountain glaciation heralds the onset of an ice age and persists for a long time (possibly tens of millions of years) after the end of an ice age. But this does not alter the fact that the peak of an ice age should be defined as the period during which continental icesheets affect the *middle latitudes*.

The importance of continental arrangements seems to come out, time and again, as a major factor in the explanation of ice ages. We have even proposed (in Chapter 1) that the reason for the lack of a Jurassic ice age may have been the lack of any large land area in a polar location. Even though there may have been global cooling at this time, there was no possibility for the build-up of high-latitude ice and the embryo ice age was never able to develop. Similar circumstances may have prevailed during those parts of Precambrian time for which ice-age traces are lacking.

At this point we can ask ourselves another question. If ice ages really do occur at intervals of 150 million years or so, can we think also in terms of global 'warm ages' which separate the periods of cold? In other words, if we can talk of the 'winters of the world' can we also talk of the 'summers of the world'?

We can test this idea by examining the periods around 75, 225, 375, 525, 675 and 825 million years ago, which would represent the half-way stages between the ice ages discussed in this book. The evidence in favour of global warming at these times is inconclusive, especially for the early part of the geological record. However, Grant Young has referred to the ambiguous evidence of global warmth in sediments interbedded with the glacial deposits of Upper Proterozoic times. We referred in Chapter 1 to the great climatic amelioration of Cambrian time which was accompanied by an evolutionary 'explosion'. About 375 million years ago, during the Devonian period, there was a vast land area in the southern hemisphere high-latitude zone, with a higher grouping of continents below 30°S than at any time since. And yet this period was apparently one of global warmth. In the area where, given favourable circumstances, large-scale polar glaciation might have been expected, there are no signs of glacial action by ice caps or icesheets. If we look at the palaeoclimatic reconstructions for the next of our possible warm episodes, about 225 million years ago, we find that this coincides with the Triassic period—a time of global warming but also a time of plant and animal extinctions on a vast scale. The cause of these extinctions is not known for certain, but there can be no doubt that global environments changed drastically following the end of the Permo-Carboniferous ice age. During the Cretaceous period, about 75 million years ago, the climate of the world was again warmer than average. The climate in western Europe at that time was tropical, and the gradual decline of temperatures since the end of the Cretaceous has been well documented from many different lines of research.

The hypothesis presented above is full of difficulties and uncertainties. Not least among the problems involved in the recognition of global warmings and coolings is the difficulty of matching up the new palaeogeographical ideas of the last decade or so with out-dated ideas in the fields of palaeoclimatology and palaeontology. The distributions of ancient climates, ancient vegetation zones, and ancient plant and animal distributions as represented by fossils have not yet been adequately reinterpreted in the light of the various continental arrangements now recognized by the geophysicists. But, for what it is worth, an outline of the main climatic episodes which appear to have affected planet Earth is shown on page 224 in a diagram. This hypothetical sequence of events is not necessarily believed by the other authors involved in this book, but it is presented with some conviction as a perfectly valid working hypothesis which is well worth testing. The 150-million-year interval between ice ages seems to be more convincing than any other interval which might be proposed on the basis of the currently available evidence. It is also more convincing than the idea that the scatter of ice ages through time can be explained as the result of chance combinations of circumstances.

A Sense of Scale

From the foregoing paragraphs it seems that, in our attempts to understand the causes and the spacing of ice ages through geological time, we should examine only those ice-age theories which are applicable on a vast timescale. In Chapter 2 we examined a number of theories which might be relevant on a scale of tens of millions of years: among these were the 'internally generated' factors such as continental drift and sea-floor spreading, mountain-building, large-scale sea-level changes due to the 'assembly' and fragmentation of supercontinents, and variations in the Earth's core leading to palaeomagnetic changes.

But none of these theories seems adequate on its own (or indeed in combination with one or more of the others) to account for the periodicity of ice ages. The important factor may, therefore, be related to astronomical changes such as alterations in the energy output of the Sun, galactic changes (page 43), or else chemical changes in the atmosphere, hydrosphere and lithosphere. There seem to be long-term oscillations in all of these, but no 150-million-year 'period' has yet been convincingly demonstrated. The 'galactic period' referred to by Rhodes Fairbridge in Chapter 5 is 200-250 million years, and this is difficult to relate to the spacing of ice ages we have been discussing. Possibly there is another galactic or cosmic period of 150 million years or so which has not yet been properly identified. Possibly there is a 150-million-year period in connection with the Earth's molten interior, leading to rapid palaeomagnetic changes which might be related to the onset of ice ages. However, so little is known about the chronology of geomagnetic reversals before the Cretaceous period that no long-term oscillations or frequencies can be identified.

For the moment, therefore, the root cause of ice ages remains elusive. All we can say is that, given 'favourable' continental and oceanic arrangements and 'favourable' alignments of high mountain ranges, ice ages seem to occur on schedule at approximately 150-million-year intervals. During the intervening warm intervals ice ages do not occur, even when continental arrangements and other factors are theoretically favourable. We are clearly involved, therefore, in

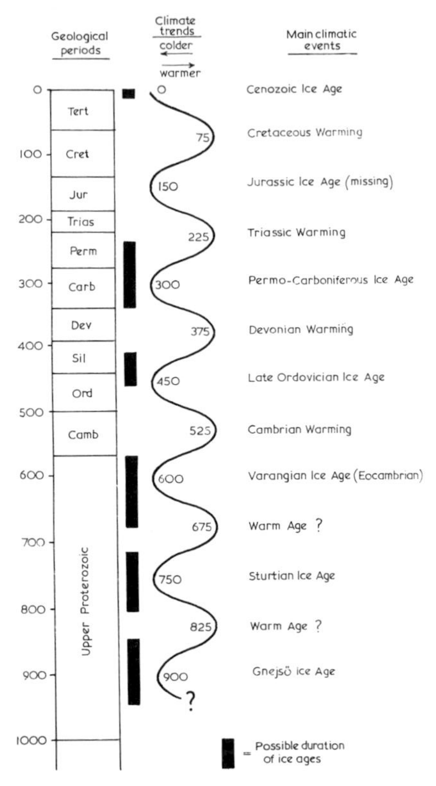

Hypothetical sequence of ice ages and 'warm ages' during the last 1,000 million years. The 150-million-year spacing of ice ages shows clearly, except for the missing ice age of Jurassic time.

the search for a mechanism of *climatic change* on a vast scale.

When we examine the climatic changes which occur *within* ice ages, our search for causes becomes somewhat easier. Oscillations on a scale of hundreds or thousands of years are still difficult to identify with any degree of certainty, but changes on a scale of tens of thousands of years can be recognized with relative ease within the later part of the present ice age. Glacial and interglacial stages seem to occur on a timescale of 100,000 years or so, and radiometric dating and other techniques have enabled Earth scientists to build up a reasonably reliable history of climatic change over the past million years or more. The main changes of climate have been described by John Andrews in Chapter 7, and he has also discussed the possible causes of these changes.

At this scale we can discount factors like continental drift and mountain-building, and the main emphasis must be on the externally generated causes discussed in Chapter 2. The spacing of glacial and interglacial stages might be controlled by changes in the solar constant (page 40), or by 'internally regulated' rises and falls in the carbon-dioxide content of the atmosphere, or by changes of sea-level or shifts in the flow of ocean currents. Alternatively, some basic instability of the atmosphere might be proposed, with self-perpetuating and self-reinforcing changes in atmospheric circulation leading to the intermittent build-up and decay of mid-latitude icesheets. However, as mentioned in several of the chapters of this book, the recognition of real cause-and-effect relationships is

CLIMATIC VARIATION	SCALE UNITS	PERIODS	TERMS
1st Order	Decades	25-100 years	Minor fluctuations, eg. short-lived glacier surges
2nd Order	Centuries	250-1000 years	Major fluctuations of postglacial and historic time, eg. 'Little Ice Age'
3rd Order	Millennia	2500-10,000 years	Stadial and interstadial episodes within glacial stages
4th Order	10,000 years	25,000-100,000 years	Glacial and interglacial stages

Orders of magnitude of climatic change which are relevant for the oscillations occurring within a single ice age. The relevant theories of climatic change at these scales probably have little to do with 'internal' factors such as mountain-building and continental drift.

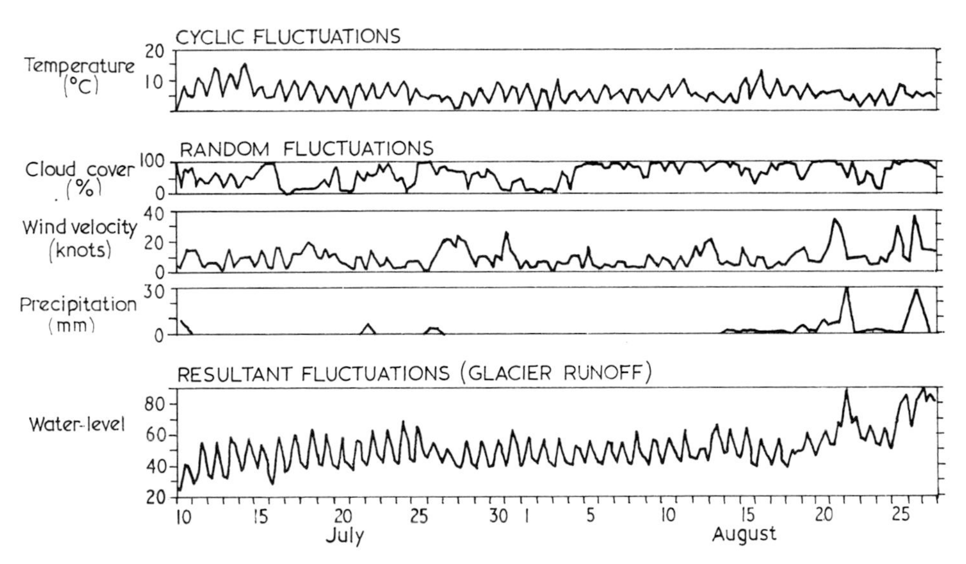

Environmental oscillations near the margin of an Icelandic ice cap. Some fluctuations are cyclic and others are random. The resultant fluctuations (in this case, of glacier runoff) show some cyclic and some random elements. The same may be true of climatic oscillations on the scale of millennia, centuries and decades.

extremely difficult. The mere 'matching up' of periodicities may lull Earth scientists into a false sense of security in the belief that the problem of ice-age climatic change has been solved.

Two groups of theories seem to be particularly useful for the understanding of climatic change in the medium term.

First there are the theories related to the behaviour of ice. These theories have not been adequately considered in the past by students of climatic change, but recent great advances in glaciology have given them added substance. The idea that icesheets regulate themselves through a complicated interplay with sea-levels, ocean temperatures and atmospheric circulations is an exciting one, and it seems to be quite valid for the explanation of glacial-interglacial oscillations.

These oscillations may also be explained, on the other hand, by reference to the 'Milankovich Model' of astronomical periodicities. As mentioned in Chapter 2, there are three main cycles in the Earth's orbit which can each affect the climate by small amounts when operating independently: when all three cycles are examined together, a recurring coincidence of phases favourable for hemispheric cooling can be recognized. A number of authors have drawn graphs of northern-hemisphere cooling and warming phases calculated on the basis of the Milankovich model. Recently the model has received considerable support from the analysis of climatic periodicities identified in studies of deep-sea cores. One paper, published late in 1976 by Hays, Imbrie and Shackleton, confirmed the reality of a 100,000 year cycle, a cycle of 42,000-43,000 years, and a cycle of 24,000 years. They concluded that 'changes in the Earth's orbital geometry are the fundamental cause of Quaternary ice ages' (by this they mean Quaternary glacial stages).

Their conclusion has been enthusiastically accepted by science writers such as Nigel Calder and John Gribbin and by many climatologists and Earth scientists also. But we should ask ourselves whether changes in the Earth's orbital geometry are really the *fundamental* causes of glacial and interglacial stages. In the first place, does the oxygen-isotope record really give a reliable indication of the expansion and contraction of the mid-latitude icesheets? This is a matter of some dispute. The record may be reliably interpreted in terms of oceanic cooling or warming, but then warmings and coolings may or may not coincide with the glacial and interglacial stages recognized on geological or glaciological criteria. If they do not (and this is quite likely), then the 'fundamental cause' of Quaternary glacial stages may not be so fundamental after all . . .

When we consider the climatic changes which have occurred within the last 100,000 years or so, a number of other periodicities can be recognized in addition to those mentioned above. These are medium- and short-term periodicities which are not of particular relevance for an understanding of ice ages through geological time, but which are relevant when we consider the planet's climatic future. Changes in solar radiation may be important here, and short-lived climatic oscillations do seem to correlate quite well with the 'sunspot cycles' of 11, 22 and 44 years. Climatic or weather shifts have been plotted on all sorts of scales from weeks to seasons, years, decades and centuries. Some of the patterns of change which have been recognized can be explained as random, while others may be related to gradual changes in atmospheric circulation. On a scale of years or decades, feedback mechanisms may operate, with such factors as northern-hemisphere snowcover, routes taken by depressions, and ocean temperatures affecting one another in complex ways. There appear to be links beween the Earth's weather patterns and changes in the magnetic field, particularly on a scale of centuries and millennia; and John Gribbin has argued strongly that the new science of 'magneto-meteorology' might provide many clues concerning the mechanisms of short-term climatic change. He has also suggested that there may be links between sunspot activity, earthquakes, volcanic eruptions and changes in the climate, and his case is a fairly convincing one. Palaeotemperature changes recorded in ice cores from Greenland indicate, in addition to an apparent 13,000-year cycle associated with major ice expansions and contractions, oscillations with intervals of 180 years, 80 years and 30-40 years. The causes of these oscillations are, however, imperfectly understood.

The most important of the medium-scale oscillations of climate appears to bring a world-wide lowering of temperature every 2,500 years or so. The latest of these cool periods was the 'Little Ice Age', which is well documented in the history of Western Europe, the Far East and other regions. The cooling was not simultaneous everywhere, but in the North Atlantic area several of the key events of history were associated with a long series of harsh winters, advancing glaciers and widespread deaths among both livestock and human populations.

In Iceland and Greenland the period 800-1,000AD was a time of warm, dry climate. Then the climate deteriorated sharply, and within 400 years the Greenland Viking settlement was exterminated by the cold and by lack of contact with the outside world. Shipping had become virtually impossible around the coasts of Greenland as a result of a huge increase in the amount of sea ice carried southwards from the Arctic.

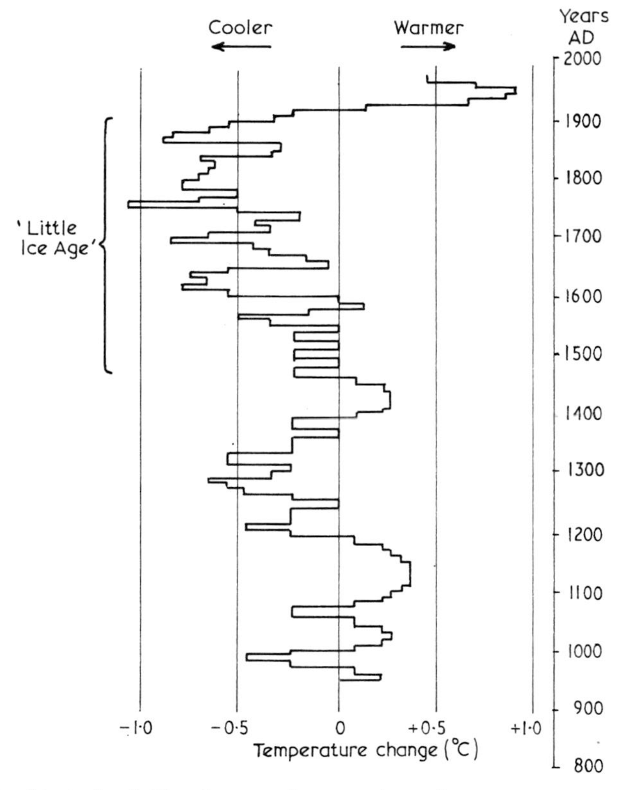

The 'Little Ice Age' in Iceland. The diagram shows estimated mean temperatures for every decade since about 950AD. The strong cooling which lasted from about 1450 until 1900 brought a time of great hardship to may parts of northern Europe.

A winter scene by Van der Keer portraying conditions during the Little Ice Age in Holland.

At the same time, the climate deteriorated also in Iceland, and the sixteenth and seventeenth centuries were times of great hardship. Between 1200 and 1800 the population of the country was halved, largely as a result of famine.

In Scandinavia also the Little Ice Age had dramatic effects, with successions of extremely harsh winters, advancing glaciers in the mountains and frequent harvest failures. The large-scale emigrations of Norwegians, Swedes and Finns to the New World during the nineteenth century may well have been caused, at least in part, by the extreme hardship brought about by the Little Ice Age.

Hardship was experienced also by the people of the Alpine valleys between 1500 and 1900, and even in the low-lying parts of Europe the effects of the climatic cooling were felt in runs of long, cold winters, frozen rivers and lakes, and harvest failures.

The Little Ice Age (which seems to have culminated about 200 years ago) was but the most recent of the medium-term climatic deteriorations. There now seems to be good evidence for earlier cold phases in northern Europe and in North America about 2,800 years ago, about 5,300 years ago and about 7,800 years ago. There were also widespread and violent oscillations of ice margins at the end of the last glaciation of Scandinavia and North America (see Chapter 7), and these may be part of the same series of medium-term climatic changes. A convincing case has been made out for a periodicity of 2,500 years or so, and there is good evidence that this periodicity ties in quite closely with pulsations in solar radiation.

Man is gradually learning how to adapt his technology to cope with conditions of extreme cold. This oil-drilling rig is at work on the north slope of Alaska.

Behold the Ice-Man Cometh

As a postscript to this book, we can speculate for a moment about the future. We have already seen that the Cenozoic ice age is by no means over, and that, if earlier ice ages are anything to go by, it should last for another 40 million years or so. We should by now have convinced ourselves that we are not living in the post-glacial period of this ice age. The ice will undoubtedly return, much to the dismay of many textbook writers who have blithely assumed that the ice age was a thing of the past to be studied with a certain amount of historical detachment. The period in which we live is a typical interglacial. There is nothing to mark it out from earlier interglacials, and it is certainly no warmer than the interglacials of the past. On the contrary, this interglacial is cooler than some earlier interglacials, and there are signs that it is also going to be shorter. This may well be typical of the interglacials which occur on the 'descending limb' of an ice age as it sinks ever deeper towards a state of global refrigeration. Luckily for us and for all the lifeforms of planet Earth, the process seems to stop before icesheets cover the tropical lands; but from the geological and palaeoclimatic evidence we can assume that this particular ice age still has some way to go before it reaches its culminating stage of cold.

The above points are all abstractions as far as human beings of today are concerned. What we all want to know is when the next cool phase is going to begin, when the next glaciation is going to start. In order to answer this question we must bear in mind the varying scales of climatic oscillation discussed earlier in this chapter.

If we look at the glacial-interglacial cycle there is no imminent danger; probably the present interglacial still has between 5,000 and 10,000 years to run.

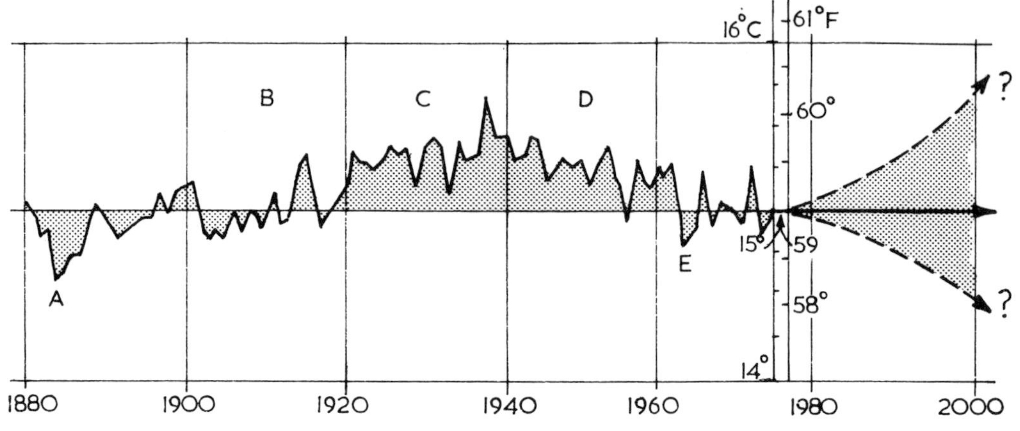

A plot of northern-hemisphere temperature changes since 1880. Although there has been a slow deterioration of climate since about 1938 there is no guarantee that it will continue to get colder.

If we look at the Neoglacial pattern of cooling and warming, the signs are, once again, that we are in no imminent danger of an onset of glaciation. The Little Ice Age has only just ended, and there is no prospect of another one, so far as we can tell, for another 2,500 years or so. Even when the next Little Ice Age does occur, we will be infinitely better prepared for it than were our ancesters of the period 1500-1900AD.

A number of authors have recently made great capital out of the threat of imminent climatic catastrophe linked with world-wide cooling and glacier expansion. For example, Nigel Calder's book *The Weather Machine and the Threat of Ice* and John Gribbin's book *The Climatic Threat* are full of dark and dire predictions about 'downward climatic spirals', weather abnormalities, and increases in the snow-covered area of the northern hemisphere. On the other hand there seems to be *no* good evidence of a substantial climatic change in progress. Even the climatic changes predicted from assorted computer models (some of them using the Milankovich 'effect' as a basis) are of such a long-term nature that they can be of little relevance over the next few thousand years. There is no adequate basis for predicting that a new widespread glaciation will be under way within the next 2,000 years. There is no adequate basis for Nigel Calder's alarmist view that a 'snowblitz' may plunge us into the next glaciation at virtually any moment. There is no adequate basis for suggesting (as John Gribbin does) that 'an ice age may be upon the world within a few hundred years, and that the immediate deterioration of the climate at the present time requires urgent attention from all responsible people'.

No one can be certain what the climatic future holds, and, while we should not accept alarmist statements uncritically, we should also be aware of the possibilities which do exist for rapid climatic change. In the previous chapter John Andrews argued strongly in favour of the idea that the climate 'flips' suddenly from the glacial mode to the interglacial mode and then back again. The sudden take-off of glacial conditions implies the setting in motion of a chain of reactions over perhaps a thousand years or so or even over a timespan as short as a few centuries. Similarly, the sudden collapse of the icesheets and the sudden surging

Signs of the next winter? Here scientists are investigating the Ross Ice Shelf, Antarctica. This area, where the shelf grades into the west Antarctic icesheet, will provide the early warnings of the disintegration of the icesheet. Mankind needs to keep an eye on it.

of icesheet segments are so rapid, in geological terms, that they should be referred to as 'geological catastrophes'.

While it is unlikely, according to the evidence currently at our disposal, that rapid shifts of climate will occur in the near future, the possibility is ever-present. Ice is, after all, a normal component of an ice age; we should expect it to make its presence felt every now and then, with or without advance warning.

BRIAN JOHN

Glossary

Ablation: Melting and other processes which reduce the mass of bodies of snow and ice.

Ablation moraine: Moraine formed by the release of detritus from glacier ice as a direct result of ablation.

Abyssal plain: Extensive plain covered by sediments on the floor of the ocean.

Accumulation: Building up of the surface of a glacier or snowfield by snowfall and other processes.

Active layer: Surface layer in a permafrost region which melts every summer.

Aeolian: Related to the activity of the wind.

Aftonian interglacial: North American interglacial which occurred between the Nebraskan and Kansan glacial stages.

Agglomerate: Mass of coarse rock fragments or angular blocks of lava produced during a volcanic eruption.

Albedo: A measure of the whiteness or brightness of a surface exposed to solar radiation.

Alluvium: Sediment laid down by running water and consisting largely of sand, silt and clay.

Alpine glaciation: Glaciation by snowfields and valley glaciers in an area of rugged mountains.

Alpine orogeny: Episode of Cenozoic mountain-building which was responsible for the creation of the European Alps.

Ammonite: Free-swimming marine mollusc with an external aragonite shell subdivided internally into chambers. Now entirely extinct. Some ammonite shells were over a metre in diameter.

Angaraland: One of the supercontinents of the Carboniferous period which later became fused with Laurasia to form the northern part of Pangaea.

Antarctic: Area within the Antarctic Circle, latitude 66½°S.

Anticline: An upfolded or arched structure in bedded rocks, generally formed as a result of pressure from the sides.

Archaean: One of the main divisions of Proterozoic (Precambrian) time.

Arctic: Area within the Arctic Circle, latitude 66½°N.

Arête: Sharp narrow ridge running between two alpine peaks.

Armorican (Hercynian) Orogeny: Mountain-building episode which occurred during the later part of the Palaeozoic era.

Atmosphere: The envelope of gases which surrounds the Earth.

Basal sliding: The sliding or slipping of a glacier over its bed, aided by a lubricating film of water.

Basalt: Dark-coloured fine-grained lava. The commonest of all lava types.

Bedding-plane: Surface which separates one layer of sedimentary rock from another.

Bergy bit: A fragment of broken floating ice derived from an iceberg.

Beringia: Ancient land area including Alaska and Northeast Siberia, now partly submerged by a rise of sea-level.

Biber glaciation: Possible early glacial stage in the European Alps (Cenozoic ice age).

Biosphere: The 'life layer' at the surface of planet Earth.

Biostratigraphy: Stratigraphy based upon differences in the fossil content of successive rock layers.

Bivalve: Mollusc which has a two-valved shell protecting the body.

Boreal forest: The northern forest of coniferous trees which stretches in a broad belt across Eurasia and North America.

Boulder-clay: The old name for till.

Brachiopod: Marine animal with a shell consisting of two valves, one of which is normally much larger than the other.

Braided stream: Stream flowing in many connected channels and usually heavily laden with debris from a glacier.

Brash-ice: Floating ice made up of broken pieces of glacier ice together with fragments of pack-ice.

Caledonian Orogeny: Mountain-building episode which occurred in Silurian and early Devonian times.

Calving: The mechanism by which icebergs and smaller ice fragments break away from a floating glacier margin.

Cambrian: The earliest geological period of the Palaeozoic. Lasted from 570 to 500 million years ago.

Carbonates: Materials made of carbonates such as calcite, dolerite or aragonite.

Carboniferous: Literally 'coal-bearing'. This period lasted from 345 to 280 million years ago.

Catastrophism: Theory, now largely discredited, that past geological changes occurred as a result of sudden great catastrophes such as Noah's Flood.

Cenozoic: The 'recent life' era which began about 65 million years ago. It includes both the Tertiary and Quaternary periods.

Chatter-marks: Small crescentic fractures on a rock surface affected by overriding ice.

Chronostratigraphy: Stratigraphy based upon the absolute or radiometric ages of successive rock layers.

Cirque: Armchair-shaped hollow in the mountains eroded by a small glacier.

CLIMAP: Climate Long Range Investigation Mapping and Prediction.

Coal: Rock formed from peat and other vegetable matter after a long period of compaction.

Coal Measures: Rock formations of Upper Carboniferous age which contain many beds of coal.

Cold ice: Glacier ice which is well beneath its pressure melting temperature.

Consolidation: Transformation of a soft loose material into a hard, dense rock.

Continent: A large landmass rising steeply from the deep ocean floor.

Continental crust: That part of the crust constituting the continents and made up largely of granitic rocks.

Continental drift: The movement of the continental blocks relative to one another across the surface of the Earth.

Continental shelf: That part of a continent which is shallowly submerged by the sea.

Convection: Transfer of heat caused by density variations in a fluid.

Coral: A hard stone of calcium carbonate, made up of the skeletons of millions of small marine animals.

Cordillera: A long mountain range formed on a continent adjacent to an ocean trench.

Core: The central sphere of the Earth, partly solid and partly liquid.

Corrie: The English name for a cirque.

Craton (Kraton): One of the old stable portions of the Earth's crust, largely unaffected by earth movements during Phanerozoic time.

Creep: The process by which glacier ice crystals adjust their positions with respect to one another, thereby allowing the ice to deform or move.

Crescentic gouges: Crescent-shaped pits or fractures in a rock surface affected by glacial erosion.

Cretaceous: The final period of the Mesozoic era, ending about 65 million years ago.

Crevasse: Deep crack in the surface of a glacier, usually formed as a result of surface tension during ice movement.

Crinoid: Marine animal of the family of echinoderms, with a stem and a body protected by calcite plates.

Cross-bedding: Bedding within a stratum which is inclined at an oblique angle to the main bedding planes.

Crust: The outermost shell of the Earth, varying from *c*.5km thick under the oceans to 60km thick under continental mountain chains.

Crustal wandering: A general term which describes changes in the arrangement and latitude of landmasses over the face of the Earth during geological time.

Cruziana: Marks made by trilobites in early marine sediments.

Cryosphere: The sphere of frost, snow and ice on the surface of the Earth.

Crystal: A chemical substance with a definite geometrical form.

Crystalline rock: Rock made up of varied mineral crystals which have formed as a result of cooling from a molten state.

Cwm: The Welsh word for a cirque or corrie.

Cyclothem: A repeated sequence of sedimentary rock layers such as those in the Carboniferous coal measures.

Dead ice: Stagnant ice (usually near a glacier snout) which is not being maintained by accumulation and glacier movement.

De Geer moraine: Moraine made up of delicate ridges of till formed one after another at or near the retreating margin of a glacier in deep water.

Deep-sea trench: Deep trench in the ocean formed where an oceanic plate is being subducted. Such trenches may be 10,000m deep.

Deflation: The removal of loose surface materials through the action of the wind.

Denudation: The wearing down of the land by the processes of weathering and erosion.

Deposition: The dumping or accumulation of materials following transport by natural agencies.

Desert: An area lacking in surface moisture, making it difficult for plants or animals to survive.

Devensian glacial stage: The last glacial stage in the British Isles, equivalent to the Weichselian of Northwest Europe.

Devonian: Geological period in the middle part of the Palaeozoic era, lasting from 395 to 345 million years ago.

Diorite: Coarse-grained intrusive igneous rock.

Dip: Angle of inclination (of sedimentary rock layers, etc.).

Dolerite: Dark-coloured igneous rock, resembling basalt but composed of larger crystals.

Dolostone: Rock made largely of dolomite, one of the carbonate minerals.

Donau glaciation: The earliest glacial stage of the 'classical' glacial sequence of the European Alps.

Drift: A broad term for all materials deposited by glacial and fluvioglacial processes.

Dropstone: Stone or boulder dropped onto the sea bed after transport by floating ice. Usually found in glacio-marine sediments.

Drumlin: A low, rounded and elongated hill made of glacial deposits and shaped by overriding ice.

DSDP: Deep-Sea Drilling Program.

Dune: Mound of sand or other fine material created by the action of the wind.

Dwyka glaciation: Name given to the Permo-Carboniferous glaication of southern Africa.

Eccentricity (of orbit): Measure of the closeness of the Earth's orbit to a circle.

Ecological niche: An environment with its own unique characteristics suitable for occupation by a specific plant or animal community.

Eemian: The last interglacial of Northwest Europe.

End moraine: Mass of glacial deposits formed near the snout or terminus of a glacier.

Englacial: Contained within a glacier.

Eocambrian ice age: Term sometimes used for the Varangian ice age which occurred just before the beginning of the Cambrian period (*c.*600 million years ago).

Eocene: One of the epochs of the Tertiary period, occurring between the Palaeocene and the Oligocene.

Eon: The longest interval of geological time, composed of several eras grouped together.

Eozoic: Term sometimes used for one of the eras of Precambrian time.

Epoch: An interval of geological time (such as the Pleistocene) which is shorter than a period. Several epochs make up a period.

Equilibrium line: The line or zone across a glacier which separates the accumulation zone from the ablation zone.

Era: An interval of geological time made up of several periods grouped together.

Erosion: The wearing away of rocks at the Earth's surface by natural agencies such as rivers or glaciers.

Erratics: Boulders or stones transported by glacier ice from their place of origin and left in an area of different rock-type.

Esker: Long winding ridge of sand and gravel, usually formed by a glacial meltwater stream in a subglacial tunnel.

Eustasy: Movement of sea-level caused by changes in the amount of water present in the world ocean.

Evaporite: A rock formed as a result of evaporation, with precipitation of minerals from a saturated solution.

Evolution: The gradual change in organisms through time, resulting in the origin of new species.

Extrusive volcanic activity: Volcanic eruption during which materials such as lava and ash are laid down on the Earth's surface.

Faceted stone: A stone with one or more flattened faces cut as a result of glacial erosion.

Facies: A rock unit or group of units having particular features which reflect some specific environmental conditions.

Fault: A fracture in the Earth's crust where one side has moved relative to the other.

Fauna: A broad term for animal life; sometimes used for either an assemblage or a community of species.

Feedback: Reaction or response where there is a cause/effect relationship between two variables.

Firn: Frozen water which is intermediate between snow and glacier ice in its crystal structure. Usually it is buried beneath a glacier surface by fresh snowfall.

Fjord (fiord): Deep glacial valley cut by an outlet glacier and now flooded by the sea.

Flandrian interglacial: The present (Holocene) interglacial of the Cenozoic ice age.

Flora: Broad term for plant life; sometimes used for either an assemblage or a community of species.

Flowtill: A till emplaced by the flowage of material on a melting ice surface.

Fluted moraine: Moraine arranged into elongated ridges (flutes) and troughs aligned parallel with the direction of ice movement.

Fluvioglacial: Pertaining to the activity of glacial meltwater flowing either within or beyond the confines of a glacier.

Fold: Flexure or bend in the rocks caused by compression in the Earth's crust.

Foliation: Banded or laminated structure in rock or glacier ice.

Foraminifera: Small marine organisms with jelly-like bodies surrounded by a shell or casing of carbonate of lime.

Forebulge: Area of raised crust beyond an icesheet margin which compensates for the isostatic depression of the ice-covered area.

Formation: The fundamental unit in the subdivision of rock sequences. Usually a rock formation has a distinctive lithology.

Fossil: Anything found in a rock which is recognizable as the remains of a plant or animal.

Frost-heaving: Disturbance or raising of the ground surface as a result of the expansion of small ice crystals.

Galactic plane: Plane on which the spiral arms of the Galaxy are arranged.

GARP: Global Atmospheric Research Program.

Gastropod: Animal protected by a shell which is coiled in a spiral; the shell is not divided into chambers.

GCM: Global Circulation Model.

Geology: The study of the structure and composition of the Earth.

Geomagnetism: The magnetism of the Earth, revealed in the magnetic properties of rocks at or near the surface.

Geomorphology: The scientific study of landscape, landforms and Earth surface processes.

Geophysics: The science that uses the principles of mathematics and physics in the examination of the Earth's interior and its internal constituents.

Geosyncline: A large trough formed along the margin of a continent and affected by plate tectonics and sea-floor spreading.

Glacier: Body of ice nourished largely by snowfall and flowing downhill under the influence of gravity.

Glacial stimulus: Encouragement of development or evolution as a result of glaciation and associated changes of climate.

Glacial stress: Stress exerted on plant and animal life as a result of glaciation and associated severe or rapid changes of climate.

Glacial Theory: Theory attributing certain modifications of the Earth's surface to the work of glaciers.

Glacio-eustasy: Movement of sea-level caused by the extraction or addition of oceanic water related to the growth and decay of icesheets.

Glaciofluvial: Pertaining to glacial meltwater.

Glaciology: The scientific study of glaciers, snow and ice.

Glossopteris flora: Flora associated with the seed fern *Glossopteris*, and particularly widespread in Gondwanaland during the Permian period.

Gnejsö ice age: Term (not universally used) for the Precambrian ice age which occurred about 900 million years ago.

Gondwanaland: Ancient southern hemisphere supercontinent made up of South America, Africa, India, Antarctica and Australia.

Gowganda glaciation: Glaciation or ice age represented by the Gowganda formations of North America. The ice age is also called the Huronian, and it occurred about 2,300 million years ago.

Granite: A coarse-grained igneous intrusive rock, often white or pink in colour.

Graptolites: Extinct group of marine animals, widely used for dating rock layers of early Palaeozoic age.

Greenhouse effect: The effect of the atmosphere in permitting the passage of short-wave solar radiation but largely preventing long-wave radiation from returning into space. This allows the maintenance of relatively high temperatures at the Earth's surface.

Greenstone: Basic volcanic rock such as basalt which has been altered by metamorphism.

Gritstone: A coarse type of sandstone, composed for the most part of large and sometimes angular grains of quartz.

Groove: Deep scratch in a rock surface caused by glacial abrasion.

Ground moraine: General term for a widespread sheet of till with a low surface relief.

Günz glaciation: The second of the glacial stages recognized by Penck and Brückner in the European Alps.

Half-life: Period over which an initial quantity of a radioactive isotope such as C^{14} reduces, through decay into other isotopes, by half.

Head: Thick layer of frost-shattered rock fragments and other materials covering a bedrock slope.

Headwall: Rock cliff cut by glacial erosion and other processes at the head of a cirque glacier.

Heavy oxygen: Oxygen containing a relatively large amount of the stable isotope O^{18}.

Hercynian Orogeny: Mountain-building episode which occurred in Europe in Permo-Carboniferous times.

Holocene interglacial: The present interglacial stage (Flandrian).

Hominids: Family of primates including man and his close relatives.

Hoxnian interglacial: The last-but-one interglacial in the British Isles, thought to equate with the Holstein interglacial of Northwest Europe.

Humid climate: Climate in which there is sufficient rainfall and surface moisture to sustain a wide range of plant and animal life.

Huronian ice age: The earliest known ice age, dated to about 2,300 million years ago.

Hydrosphere: The 'sphere of water' on the Earth's surface, represented for the most part by the world ocean.

Ice age: A prolonged period (usually millions of years) during which icesheets intermittently occupy the high and middle latitudes of the planet.

Iceberg: A large detached piece of glacier ice which floats free after calving from the glacier snout.

Ice cap: Large glacier usually found in a mountainous or plateau area and not greatly affected by the relief of the land surface which it covers.

Ice-contact structures: Slumping and other structures formed in sediments adjacent to melting glacier ice.

Ice rafting: The transport of glacial debris by floating ice such as pack ice floes and icebergs.

Icesheet: A vast glacier of dome shape and more than 50,000 square kilometres in extent. It rests upon rock which may be partly depressed beneath sea-level by the weight of ice.

Ice shelf: A large floating glacier formed in a coastal embayment and maintained for the most part by snowfall on its flat upper surface.

Ice wedge: Wedge of ground ice filling a vertical crack in the surface of the permafrost.

IDOE: International Decade of Ocean Exploration.

Igneous rock: Rock formed by the solidification of molten material from the Earth's interior.

Illinoian glaciation: The last but one glacial stage in North America, thought to be the equivalent of the European Saalian.

Infracambrian ice ages: The ice ages which preceded the Eocambrian or Varangian ice age. The two Infracambrian ice ages are also labelled Gnejsö (*c.*900 million years ago) and Sturtian (*c.*750 million years ago).

Interglacial: Part of an ice age during which there is relatively little ice present on the Earth's surface.

Interstadial: Interval during a glacial stage when the climate improves and the ice retreats for a limited time.

Intrusive rocks: Rocks which have been forced in a molten state into the cracks or cavities in pre-existing rock strata.

Involutions: Complicated contortions which disturb the layering of deposits, usually indicative of periglacial conditions.

Ipswichian interglacial: The last interglacial in the British Isles (=Eemian).

Isostasy: Condition of equilibrium in the Earth's crust. Loading by ice causes the crust to sink, and unloading causes it to rise again.

Isthmus: A narrow neck of land connecting two larger land-masses.

Joint: A line of breakage or fracture cutting across the bedding or other structures of a rock.

Jökulhlaup: A sudden violent flood of meltwater resulting from the outburst of a subglacial lake or from a period of exceptionally rapid ice melt.

Jurassic: The middle period of the Mesozoic era. The time of dinosaurs.

Kame: Ridge or mound of sand and gravel formed at the edge of a wasting glacier.

Kame terrace: Terrace of sand and gravel formed between a glacier margin and a valley side.

Kansan glaciation: One of the glacial stages of North America, preceded by the Aftonian interglacial and followed by the Yarmouth interglacial.

Katabatic wind: Wind caused by the downslope drainage of cold air, for example, in a mountain valley or on the sides of an icecap or icesheet.

Kettle-hole: A pit or hollow in glacial or fluvioglacial deposits resulting from the melting of a buried mass of glacier ice.

Knock and lochan topography: Landscape of ice-scoured hollows (often containing small lakes) and upstanding rocky hillocks resulting from widespread glacial erosion or areal scouring.

Kraton (shield): See *Craton*.

Lacustrine: Formed in a lake.

Laminar flow: Movement of a fluid in which there is very little turbulent mixing.

Land bridge: Area of sea-floor exposed at a time of low sea-level and allowing the migration of plants and animals from one landmass to another.

Lateral moraine: Moraine formed along the edge of a glacier, usually in a glacial valley.

Laurasia: Northern hemisphere supercontinent which consisted of North America, Greenland and Europe west of the Urals.

Laurentide icesheet: Cenozoic icesheet covering the northern parts of the North American continent.

Lava: Molten igneous rock which has flowed from the Earth's interior to cool and solidify at the surface.

Lead: A 'channel' of open water between two areas of pack ice or between individual ice floes.

Limestone: Rock rich in calcium carbonate. Usually formed in clear, warm water from the remains of marine organisms such as coral and sea shells.

Lithology: The term used for the physical characteristics of a rock layer.

Lithosphere: The outer rigid shell of the Earth, made up for the most part of non-organic rocks.

Lithostratigraphy: Stratigraphy based on the differing physical characteristics of successive beds of rock.

Little Ice Age: Episode of cool climate and glacier advances in the northern hemisphere which lasted from the Middle Ages until the early part of the present century.

Load: The detrital material in transport by a river, glacier or other agency. Alternatively, the material responsible for the isostatic depression of part of the Earth's crust.

Lodgement till: Till deposited by 'plastering' on the bed of a glacier by active ice at or near its pressure melting point.

Loess: Fine-grained windblown sediment originally of glacial or fluvioglacial origin.

Magma: The molten material from the Earth's interior and from which all igneous rocks are made.

Magnetic epoch: Period of time during which the Earth's magnetic field was either consistently normal or consistently reversed.

Magnetic reversal: Switch of the Earth's magnetism from north to south or vice versa.

Mantle: That part of the Earth's interior beneath the crust and above the core. It behaves largely as a solid, but is molten in its upper layers.

Marine or maritime icesheet: An icesheet which rests upon bedrock which is partly or wholly below sea-level, making it very unstable.

Medial moraine: Ridge or 'stripe' of moraine within a glacier and parallel with the direction of ice flow.

Meltout till: Till formed on or beneath a glacier by the 'release' of detritus during ice melting.

Meltwater: Water formed as a result of the melting of snow or glacier ice.

Mesozoic: The 'middle life' era, which lasted from 225 until 65 million years ago.

Metamorphic rock: Rock which has been greatly changed by heat or pressure.

Metazoan: Animal whose body is formed of many cells.

Meteorite: Solid body that has fallen to the Earth's surface from an extra-terrestrial source.

Mid-oceanic ridge: Elongated submarine elevation marking the position of a constructive plate margin.

Milankovich Theory: Theory of climatic change related to variations in the amount of solar radiation received at different latitudes and influenced by variations in celestial mechanics.

Mindel glaciation: The second of the 'classical' sequence of glacial stages recognized in the European Alps.

Mineral: A solid inorganic substance which has its own recognizable physical and chemical properties.

Miocene: One of the epochs of the Tertiary period.

Mississippian: In North America the early part of the Carboniferous period.

Mixtite: Sediment made up of a mixture of stones, gravel and finer materials with few traces of sorting or layering. The term is used where the genesis of the material is in doubt.

Moraine: Ridge or mound of glacial deposits formed at or near an ice margin.

Moulin: Deep pit through which surface meltwater descends into the body of a glacier.

Mudflow: Mass movement of saturated fine-grained material and stones on a slope, often occurring quite rapidly.

Mudstone: A fine-grained sedimentary rock, usually formed of silt and clay laid down in deep water.

Muskeg: Type of northern bog vegetation consisting of mosses, sedges and intermittent trees.

Nautiloids: Marine animals with external shells of aragonite which are divided into chambers.

Nebraskan glaciation: The earliest of the North American Pleistocene glacial stages.

Neoglacial: Period of glacier growth during the present interglacial. Started about 500 years ago.

Névé: See *Firn.*

Nivation: Process of erosion by which snow-patches enlarge the rock hollows in which they rest.

NORPAX: North Pacific Experiment.

Nunatak: Hill or mountain summit projecting through the surface of a glacier.

Obliquity (of ecliptic): The tilt angle of the Earth's axis of rotation.

Ocean basin: Basin contained by continental margins and made up of mid-ocean ridges, abyssal plains, deep-sea trenches and other elements.

Oceanic crust: Crust making up the ocean floors and generated at constructive plate margins along mid-oceanic ridges.

Old Red Sandstone: The name given in Britain to the Devonian period of 395 to 345 million years ago.

Oligocene: One of the epochs of the Tertiary period.

Ordovician: A period of the Lower Palaeozoic era which lasted from 500 to 430 million years ago.

Orogeny: An episode of mountain-building.

Ostracods: Small crustaceans with bivalved shells of calcium carbonate. There are both fresh-water and marine forms, and many genera still survive.

Outlet glacier: Large glacier carrying ice from an ice cap or icesheet towards the coast.

Outwash stream: Stream of glacial meltwater originating on, within or beneath a glacier and flowing beyond the glacier snout.

Pack ice: Floating ice which is not fixed to the shore. Usually the ice is broken into slabs or floes.

Palaeocene: The oldest epoch of the Tertiary period, which began about 65 million years ago.

Palaeoclimatology: Science of ancient climates.

Palaeogeography: Science of ancient geography involving the reconstruction of old landscapes and continental positions from widely varied evidence.

Palaeontology: The study of ancient organisms, mostly based upon the analysis of plant and animal fossils.

Palaeozoic: The 'ancient life' era, lasting from about 750 to 225 million years ago.

Palsa: Mound of peat or other deposits formed through the expansion of small lenses of ground ice.

Palynology: The study of fossil spores, especially pollen.

Pangaea: The largest supercontinent of Palaeozoic time, formed by the 'fusing' together of Gondwanaland, Laurasia and Angaraland.

Patterned ground: Ground with polygons, circles, stripes or other markings often associated with the presence of permafrost.

Periglacial: Characterized by frost-shattering and other cold-climate processes, with or without the presence of permafrost.

Period: One of the main subdivisions of an era of geological time.

Permafrost: Permanently frozen ground which may be hundreds of metres thick.

Permian: The last period of the Palaeozoic era, ending about 225 million years ago.

Permo-Carboniferous ice age: Ice Age of late Palaeozoic times which affected Gondwanaland in particular.

Petrology: The scientific study of rocks, including mineralogy, petrography, etc.

Phanerozoic: That part of geological time which has elapsed since the beginning of the Cambrian period. Includes the Palaeozoic, Mesozoic and Cenozoic eras.

Pingo: Large mound caused by the growth of a single ice lens beneath the ground surface.

Plankton: Organisms that float and drift passively in the water. Phytoplankton are plant forms and Zooplankton are animals.

Plastic deformation: Deformation of glacier ice as a result of slight changes in the shape of crystals and adjustments of crystal arrangements, occurring especially when ice is close to the pressure melting point.

Plate tectonics: Deformations of the crust as a result of the movements of plates across the Earth's surface. Most tectonics are related to collisions between plate margins.

Plateau: An area of upland which has its surface at a more or less constant altitude.

Pleistocene: The first epoch of the Quaternary period. This is the

	epoch in which we now find ourselves.
Pliocene:	The last epoch of the Tertiary period. Ended about 2 million years ago.
Plucking:	Process by which large blocks of bedrock are dragged away by moving glacier ice.
Polar desert:	Area within the polar regions where lack of moisture and extreme cold inhibit plant and animal life.
Precambrian:	That part of geological time which began with the formation of the Earth and ended about 750 million years ago.
Precession (of equinoxes):	Gradual change in the season of closest approach to the Sun.
Pressure melting:	Melting of ice at or near a glacier bed as a result of pressure, even where temperatures are sub-zero.
Proterozoic:	Division of geological time preceding the Phanerozoic. Sometimes the word is used as a synonym for Precambrian.
Pyroclastic rock:	Igneous rock formed as a result of violent volcanic explosions. May be composed of agglomerates and tuffs.
Quartzite:	A sedimentary or metamorphic rock made up largely of quartz grains or crystals.
Quaternary:	The most recent period of the Cenozoic era. This period has not yet ended.
Radial flow:	Flow of ice outwards in all directions from a single ice cap or icesheet centre.
Radiation:	Increase through time in the diversity, abundance and geographical range of an organism.
Radiocarbon dating:	Dating based upon the measurement of the radioactive isotope carbon-14 in fossil organisms.
Radiometric age:	Age calculated on the basis of the measurement of radioactive isotopes and their decay products in rocks or fossil organic materials.
Radio-echo sounding:	Depth sounding of ice or other materials based upon the transmission of radio waves.
Ravinement:	Cutting of deep ravines or channels in bedrock by ice and meltwater torrents.
Red beds:	Red-coloured sedimentary rocks formed in a highly oxidizing environment where ferric iron is present.
Refugia:	Unglaciated areas where plants and animals are thought to survive during episodes of widespread ice cover.
Regression:	Overall fall of sea-level relative to the land.
Relief:	Difference in height between the high and low points of a region.
Rhythmite:	Sediment which shows evidence of a rhythmic or

repeated pattern of deposition.

Rime ice: Ice formed by the direct freezing of water droplets from the atmosphere on contact with cold surfaces.

Ripplemarks: Traces of ripples caused by water or wind flow across a surface of fine-grained material such as sand.

Riss glaciation: The third of the 'classical' sequence of glacial stages recognized in the European Alps.

Roche moutonnée: Rock hummock smoothed on one side and steepened on the other by the action of overriding glacier ice.

Rogen moraine: Type of moraine in which there are large elongated ridges perpendicular to the direction of ice flow. These ridges probably form subglacially.

Sandstone: Rock composed mainly of quartz grains and representing a compacted layer of sand.

Sandur (pl. Sandar): Extensive plain of fluvioglacial outwash deposits.

Sangamon interglacial: The last interglacial of the North American Pleistocene sequence.

Scree: See *Talus*.

Sea-floor spreading: The widening of oceans as a result of the generation of new sea-floor along a mid-oceanic ridge. The plates on either side are constantly moving away from the ridge crest.

Sea ice: Ice formed through the direct freezing of the sea surface.

Sedimentary rock: Rock composed of organic or inorganic sediments laid down directly after erosion and transport by the agents of denudation.

Sérac: An isolated pillar of glacier ice bounded by intersecting crevasses, for example in an icefall.

Shear plane: Plane inside a glacier along which one mass of ice moves relative to another.

Sheet-flow: Movement by ice which is relatively constant over a wide area.

Shield: A large rigid continental area made up largely of Precambrian rocks. See *Craton*.

Siltstone: Rock made through the compaction of a layer of silt and fine sand.

Silurian: One of the periods of the Lower Palaeozoic era.

Snowline: Line separating the area in which snow disappears in summer from the higher area where snow may remain throughout the year.

Solar constant: The average intensity of solar radiation received on a flat surface at right-angles to the solar beam.

Snowblitz: Theory concerning the initiation of a glaciation as a result of an extensive mid-latitude snowcover with consequent 'positive feedback' effects.

Solidification: Hardening of molten igneous materials into layers of solid rock.

Solifluxion: Slow downslope movement of soil or rock debris lubricated by water.

Speleothem: Precipitate of calcium carbonate, usually in a limestone cave.

SST: Sea surface temperature.

Stadial: Interval of glacier advance during a glacial stage.

Stage: One of the major time intervals of a geological epoch, e.g., the glacials and interglacials.

Stone polygons: Polygonal arrangements of stones at the ground surface caused by frost-heaving and related processes.

Stratigraphy: The description and classification of rock layers and their correlation with one another.

Stratum: A single layer of rock, usually of sedimentary origin.

Streamflow: Movement by a relatively narrow stream of ice which is faster than the movement of ice on either side.

Striae: Scratches caused by the abrasive action of moving glacier ice.

Stromatolites: Rocks made of laminated layers of calcium carbonate and occurring for the most part in Precambrian rock sequences. Probably formed by primitive marine algae.

Sturtian ice age: One of the Precambrian ice ages. It is named after the Sturtian tillite sequence of Australia, and is dated to about 750 million years ago.

Subglacial: Beneath a glacier.

Sublimation: Phase change of snow or ice into water vapour.

Subduction: The dragging down of a plate margin where two plates are in collision. Crustal material is consumed in the subduction zone, and the collision contact is often marked by a deep-sea trench.

Supercontinent: A vast landmass formed through the coalescence or fusing together of a number of different continents.

Supraglacial: On the surface of a glacier.

Surge: Abnormally rapid movement of a glacier or part of a glacier, sometimes as a result of excessive melting on the glacier bed.

Surge Theory: Theory which states that ice ages or glacial stages may be 'triggered' by large-scale surges involving icesheets.

Syncline: Trough or downfold in the rocks, usually produced as a result of pressure exerted from the sides.

System: A sequence of strata laid down during a single geological period.

Taiga: The northern coniferous or boreal forest.

Talchir glaciation: Name used for the Permo-Carboniferous glaciation of the Indian subcontinent.

Talus (scree): Bank of frost-shattered rock fragments at the foot of a steep slope.

Tectonic: Pertaining to crustal movement, deformation or uplift.

Terminal moraine: Moraine formed at a glacier snout at the position of its greatest advance.

Terrace: A flat or gently sloping surface of erosion or deposition related to a former sea-level or river-level.

Tertiary: The 'third life' period lasting from 65 to 2 million years ago. Part of the Cenozoic era.

Tests: Fossil bodies of small marine organisms such as foraminifera.

Tethys: That part of the Late Palaeozoic world ocean bounded by Laurasia in the north and northwest and Gondwanaland in the south and southwest.

Thermokarst: Pitted terrain formed as a result of the melting out of ground ice masses in permafrost areas.

Tidal current: Current in the sea created by the rise or fall of the tide.

Till: Deposit laid down directly by glacier ice.

Tillite: Solid rock formed by the compaction of till over a long period of time.

Tor: Upstanding rocky mass on a rounded hill slope or summit.

Transgression: Overall rise of sea-level relative to the land.

Triassic: The first period of the Mesozoic era. Started about 225 million years ago.

Trigger mechanism: Mechanism which sets in motion a chain of reactions culminating in an ice age or glacial stage.

Trilobite: Extinct aquatic animal with an oval flattened and segmented body divided into three sections. Trilobites are especially valuable for the correlation of Lower Palaeozoic sedimentary rocks.

Tuff: Rock composed of fine-grained materials (often ash) ejected during a volcanic eruption.

Tundra: High-latitude vegetation belt with mosses, lichens, sedges and dwarf trees.

Turbidite: Deposit formed on the sea bed from materials transported by a turbidity current.

Turbidity current: A flow of dense sediment and water, usually on a steep submarine slope.

Unconformity: Contact along which a younger group of rocks rests upon the eroded surface of an older group of rocks.

Uniformitarianism: Theory that all geological changes can be explained as a result of the slow operation of processes which can still be observed today.

Uplift: Raising of the land surface by tectonic or isostatic forces.

Valley glacier: Glacier confined within the walls of a valley or trough which may itself have been created or at least deepened by glacial erosion.

Valley train: Area of outwash deposits confined between the walls of a glacial trough.

Varangian ice age: Ice age which occurred just before the end of the Precambrian. Also called the Eocambrian ice age. Named after the Varanger region of Norway.

Varves: Fine-grained layered sediments formed for the most part in glacial lakes. The layering can be interpreted as resulting from annual or seasonal rhythms of deposition.

Vein ice: Ground ice in a permafrost area which is arranged in vertical or steeply inclined 'veins'.

Volcano: A fissure or vent on the Earth's surface through which material from the Earth's interior is extruded or ejected.

Vulcanology: Science concerned with the study of volcanoes and the effects of volcanic eruptions.

Warm ice: Ice which is close to its pressure melting point, allowing the creation of meltwater.

Water table: Level in the ground beneath which all pore-spaces are filled with water.

Weathering: The decomposition or breakdown of rock as a result of physical or chemical processes.

Weichselian stage: The last glacial stage in northwest Europe (=Wisconsin of North America).

Windscoop: Deep trench in the snow around the edges of a prominent object, caused by wind eddying.

Wisconsin stage: The last glacial stage in North America.

World ocean: The ocean area of the whole world, usually subdivided into smaller oceans.

Würm stage: The last glacial stage of the 'classical' sequence of the European Alps.

Yarmouth interglacial: North American interglacial which occurred between the Illinoian and Kansan glacial stages.

Zonation: Arrangement into zones or belts.

Further Reading

The following books, booklets and articles are for the most part non-technical. They are arranged chapter by chapter, and they have been chosen for those who wish to follow up particular topics in more detail. Most of the books mentioned have lists of detailed references.

Chapter 1. Planet Earth and its Seasons of Cold

CALDER, N., 1972: *Restless Earth*, BBC, London, 152pp.

COUSTEAU, J., 1973: *The White Caps*, Angus & Robertson, Brighton, 144pp.

DYSON, J. L., 1963: *The World of Ice*, Cresset Press, London, 292pp.

EICHER, D., 1968: *Geologic Time*, Prentice-Hall, N. J., 150pp.

FUCHS, V. (ed), 1977: *Forces of Nature*, Thames & Hudson, London, 303pp.

HALLAM, A. (ed), 1977: *Planet Earth*, Elsevier-Phaidon, Oxford, 320pp.

JOHN, B. S., 1977: *The Ice Age*, Collins, London, 254pp.

— —, 1979: *The World of Ice*, Orbis, London, 120pp.

KURTÉN, B., 1972: *The Ice Age*, Rupert Hart-Davis, London, 179pp.

POST, A., and LA CHAPELLE, E. R., 1971: *Glacier Ice*, Univ. of Washington Press, 110pp.

SEYFERT, C. K., and SIRKIN, L. A., 1973: *Earth History and Plate Tectonics*, Harper & Row, N.Y., 504pp.

TARLING, D. H., and TARLING, M. P., 1971: *Continental Drift*, G. Bell, London, 112pp.

WRIGHT, A. E., and MOSELEY, F. (eds), 1975: *Ice Ages: ancient and modern*, Seel House Press, Liverpool, 320pp.

Chapter 2. Ice Ages: A Search for Reasons

BEATY, C. B., 1978: 'The causes of glaciation', *American Scientist* 66, pp452-9.

BROOKS, C. E. P., 1949: *Climate through the Ages*, Benn, London, 395pp.

CALDER, N., 1974: *The Weather Machine and the Threat of Ice*, BBC, London, 143pp.

FAIRBRIDGE, R. W. (ed), 1967: *The Encyclopaedia of Atmospheric Sciences and Astrogeology*, Reinhold, N.Y., 1,200pp.

GOUDIE, A. S., 1979: *Environmental Change*, O.U.P., Oxford, 244pp.

GRIBBIN, J. (ed), 1978: *Climatic Change*, Cambridge Univ. Press, Cambridge, 280pp.

LADURIE, E. LE ROY, 1971: *Times of Feast, Times of Famine*, Doubleday, N.Y., 428pp.

LAMB, H. H., 1966: *The Changing Climate*, Methuen, London, 236pp.

LOCKWOOD, J. G., 1979: *Causes of Climate*, Edward Arnold, London, 288pp.

NAIRN, A. E. M. (ed), 1961: *Descriptive palaeoclimatology*, Interscience, N.Y., 380pp.

SCHWARTZBACH, M., 1963: *Climates of the Past*, van Nostrand, N.Y., 328pp.

Chapter 3. Glaciers and Environment

ANDREWS, J. T., 1975: *Glacial Systems*, Duxbury Press, Mass., 191pp.

ARMSTRONG, T., ROBERTS, B. and SWITHINBANK, C., 1966: *Illustrated glossary of snow and ice*, S.P.R.I., Cambridge, 60pp.

BAIRD, P., 1964: *The Polar World*, Longmans, London, 328pp.

BIRD, J. B., 1967: *The Physiography of Arctic Canada*, Johns Hopkins Press, Baltimore, 336pp.
COATES, D. R. (ed), 1974: *Glacial geomorphology*, State Univ. of N.Y., 398pp.
EMBLETON, C. and KING, C. A. M.. 1975: *Glacial geomorphology*, Edward Arnold, London, 573pp.
— —, 1975: *Periglacial geomorphology*, Edward Arnold, London, 203pp.
EVANS, J. G., 1978: *An introduction to environmental archaeology*, Elek, London, 168pp.
FAIRBRIDGE, R. W. (ed), 1968: *Encyclopaedia of geomorphology*, Reinhold, N.Y., 1,295pp.
FLINT, R. F., 1971: *Glacial and Quaternary Geology*, Wiley, N.Y., 892pp.
GOLDTHWAIT, R. P. (ed), 1971: *Till, a symposium*, Ohio State Univ., 402pp.
— —, 1975: *Glacial Deposits*, Dowden, Hutchinson & Ross, Pennsylvania, 464pp.
HAMELIN, L. E. and COOK, F. A., 1967: *Illustrated glossary of periglacial phenomena*, Université Laval, Quebec, 237pp.
HATHERTON, T. (ed), 1965: *Antarctica*, Methuen, London, 511pp.
IVES, J. D. and BARRY, R. G. (eds), 1974: *Arctic and Alpine Environments*, Methuen, London, 1,024pp.
JOHN, B. S. and SUGDEN, D. E., 1975: 'Coastal geomorphology of high latitudes', *Progress in Geography* 7, Edward Arnold, London, pp53-132.
KING, H. G. R., 1969: *The Antarctic*, Blandford, London, 276pp.
SHARP, R. P., 1960: *Glaciers*, Univ. of Oregon Press, 78pp.
SUGDEN, D. E. and JOHN, B. S., 1976: *Glaciers and Landscape: a geomorphological approach*, Edward Arnold, London, 376pp.
THORÉN R., 1969: *Picture Atlas of the Arctic*, Elsevier, Amsterdam, 449pp.
WASHBURN, A. L., 1973: *Periglacial processes and environments*, Edward Arnold, London, 320pp.

Chapter 4. The Earliest Ice Ages: Precambrian

GLEN, W., 1975: *Continental Drift and Plate Tectonics*, Charles E. Merrill, Columbus, 188pp.
HARLAND, W. B., 1964: 'Evidence of late Precambrian glaciation and its significance', in Nairn, A. E. M. (ed) *Problems in Palaeoclimatology*, Interscience, N.Y., pp119-49.
JOPLING, A. V. and McDONALD, B. C. (eds), 1975: *Glaciofluvial and glaciolacustrine sedimentation*, Soc. Econ. Paleont. and Mineral., Tulsa, Oklahoma, 220pp.
SALOP, L. J., 1977: *Precambrian of the northern hemisphere and general features of early geological evolution*, Elsevier, Amsterdam, 378pp.
SCHERMERHORN, L. J., 1974: 'Late Precambrian mixtites: glacial and/or nonglacial', *Amer. Journ. Science* 274, pp673-824.
SPENCER A. M., 1971: *Late Precambrian glaciation in Scotland*, Geol. Soc., London, Memoir 6, 98pp.
UYEDA, S., 1978: *The new view of the Earth: moving continents and moving oceans*, Freeman, San Francisco, 217pp.
WINDLEY, B. F., 1977: *The evolving continents*, John Wiley, London, 385pp.
YOUNG, G. M., 1970: 'An extensive early Proterozoic glaciation in North America', *Palaeogeog., Palaeoclimatol., Palaeoecol.* 7, pp85-101.

Chapter 5. Traces from the Desert: Ordovician

ALLEN, P., 1975: 'Ordovician glacials of the central Sahara', in Wright, A. E. and Moseley, F. (eds) *Ice Ages: ancient and modern*, Seel House Press, Liverpool, pp275-86.
BEUF, S., BIJU-DUVAL, B., CHARPAL, O. DE, ROGNON, P., GARIEL, O. and BENNACEF, A., 1971: *Les grés du Paléozoique inferieur au Sahara*, Inst. Franc. Pétrol., 464pp.
FAIRBRIDGE, R. W., 1974: 'Glacial grooves and periglacial features/in the Saharan Ordovician', in Coates, D. R. (ed) *Glacial Geomorphology*, State Univ., New York, pp317-27.
HARLAND, W. B., 1972: 'The Ordovician ice age', *Geol. Mag.* 109, pp451-6.
SPJELDNAES, N., 1961: 'Ordovician climatic zones', *Norsk.* geol. Tidsskr. 41, pp45-77.
WILLIAMS, G. E., 1975: 'Late Precambrian glacial climate and the earth's obliquity', *Geol. Mag.* 112, pp441-65.

Chapter 6. The Great Ice Age: Permo-Carboniferous

ADIE, R. J., 1975: 'Permo-Carboniferous glaciation of the Southern Hemisphere', in Wright, A. E. and Moseley, F. (eds) *Ice Ages: ancient and modern*, Seel House Press, Liverpool, pp275-86.

CROWELL, J. C. and FRAKES, L. A., 1970: 'Phanerozoic glaciation and the causes of ice ages', *Amer. Journ. Science* 268, pp193-224.

— —, 1975: 'The Late Palaeozoic glaciation', in Campbell, K. S. W. (ed) *Gondwana geology*, ANU, Canberra, pp313-31.

GIRDLER, R. W., 1964: 'The palaeomagnetic latitudes of possible ancient glaciations', in Nairn, A. E. M. (ed) *Problems in Palaeoclimatology*, Interscience, N.Y., pp155-78.

HALLAM, A. (ed), 1973: *Atlas of palaeobiogeography*, Elsevier, Amsterdam.

HARLAND, W. B. and HEROD, K. N., 1975: 'Glaciations through time', in Wright, A. E. and Moseley, F. (eds) *Ice Ages: ancient and modern*, Seel House Press, Liverpool, pp275-86.

PLUMSTEAD, E. P., 1973: 'The enigmatic *Glossopteris* flora and uniformitarianism', in Tarling, D. H. and Tuncorn, S. K. (eds) *Implications of Continental Drift to the Earth Sciences*, Vol. 1, Academic Press, London, pp413-24.

TARLING, D. H., 1978: 'The geological-geophysical framework of ice ages', in Gribbin, J. (ed) *Climatic Change*, Cambridge Univ. Press, pp3-24.

THERON, J. N. and BLIGNAULT, H. J., 1975: 'A model for the sedimentation of the Dwyka glacials in the southwestern Cape', in Campbell, K. S. W. (ed), *Gondwana geology*, ANU, Canberra, pp347-56.

Chapter 7. The Present Ice Age: Cenozoic

ANDREWS, J. T., 1973: 'The Wisconsin Laurentide ice sheet; dispersed centres, problems of rates of retreat, and climatic implications', *Arctic and Alpine Res.*, 5, pp185-99.

BOWEN, D. Q., 1978: *Quaternary Geology*, Pergamon, Oxford, 221pp.

BISHOP, W. W., and MILLER, J. A., 1972: *Calibration of Hominid Evolution*, Edinburgh.

DAY, M. H., 1969: *Fossil Man*, Hamlyn, London, 159pp.

FAIRBRIDGE, R. W., 1972: 'Climatology of a glacial cycle', *Quaternary Res.*, 2, pp283-320.

FLINT, R. F., 1971: *Glacial and Quaternary Geology*, Wiley, New York, 892pp.

FRENCH, H. M., 1976: *The Periglacial Environment*, Longman, London, 309pp.

HOWELL, F. C., 1971: *Early Man*, Time-Life, N.Y., 200pp.

KURTÉN, B., 1968: *Pleistocene Mammals of Europe*, Aldine, Chicago, 317pp.

IVES, J. D. and BARRY, R. G. (eds), 1974: *Arctic and Alpine Environments*, Methuen, London, 1,024pp.

MAHANEY, E. C. (ed), 1976: *Quaternary Stratigraphy of North America*, Strondsberg, Penn.

McINTYRE, A., KIPP, N. G. *et al.*, 1976: 'Glacial North Atlantic 18,000 years ago: a CLIMAP reconstruction'. *Geol. Soc. Amer. Mem.* 143, pp43-76.

RANKAMA, K. (ed), 1965: *The Quaternary* (Vol. 1), Interscience, N.Y., 300pp.

SHOTTON, F. W. (ed), 1977: *British Quaternary Studies*, Clarendon Press, Oxford, 298pp.

SPARKS, B. W. and WEST, R. G., 1972: *The Ice Age in Britain*, Methuen, London, 302pp.

SUGDEN, D. E., 1977: 'Reconstruction of the morphology, dynamics and thermal characteristics of the Laurentide ice sheet at its maximum', *Arctic and Alpine Res.* 9, pp21-47.

TUREKIAN, K. K. (ed), 1971: *Late Cenozoic glacial ages*, Yale Univ. Press, 606pp.

WEST, R. G., 1968: *Pleistocene geology and biology*, Longman, London, 377pp.

WRIGHT, H. E. and FREY, D. G. (eds), 1965: *The Quaternary of the United States*, Princeton Univ. Press, 922pp.

Chapter 8. Rhyme, Reason and Prediction

BRYSON, R. A. and MURRAY, T. J., 1977: *Climates of Hunger*, Univ. of Wisconsin Press, USA, 171pp.
BUDYKO, M. I., 1974: *Climate and Life*, Academic Press, N.Y., 508pp.
CALDER, N., 1974: *The weather machine and the threat of ice*, BBC, London, 143pp.
CLAIBORNE, R., 1973: *Climate, Man and History*, Angus & Robertson, London, 444pp.
COLEMAN, A. P., 1926: *Ice Ages, recent and ancient*, Macmillan, London, 296pp.
FAIRBRIDGE, R. W. (ed), 1967: *Encyclopaedia of Atmospheric Sciences and Astrogeology*, van Nostrand, Reinhold, N.Y., 1,200pp.
FRITTS, H. C., 1976: *Tree Rings and Climate*, Academic Press, London, 567pp.
GRIBBIN, J., 1976: *Forecasts, famines and freezes*, Wildwood House, London, 132pp.
— —, 1978: *The Climatic Threat*, Fontana, London, 206pp.
— —, 1978: 'Astronomical influences', in Gribbin, J. (ed) *Climatic Change*, Cambridge Univ. Press, pp133-54.
— — and LAMB, H. H., 1968: 'Climatic change in historical times', in Gribbin, J. (ed) *Climatic Change*, Cambridge Univ. Press, pp68-82.
— — and PLAGEMANN, S. H., 1974: *The Jupiter Effect*, Macmillan, London.
LAMB, H. H., 1972: *Climate: present, past and future*, Vol. 1, Methuen, London, 648pp.
— —, 1977: *Climate: present, past and future*, Vol. 2, Methuen, London, 866pp.
MATTHEWS, S. W., 1976: 'What's happening to our climate?' *Nat. Geog. Mag.* 150(5), p.576.
NAS, 1975: *Understanding Climatic Change*, Washington, 239pp.
SINGER, S. F., 1975: *The Changing Global Environment*, D. Reidel, Dordrecht.
SCHNEIDER, S. H. (with L. E. MESIROW), 1976: *The Genesis strategy: climate and global survival*, Plenum, N.Y., 419pp.
VELIKOVSKY, I., 1973: *Earth in Upheaval*, Abacus, London, 263pp.

Index